# Universitext

*Advisers*

F. W. Gehring
P. R. Halmos
C. C. Moore

Elmer G. Rees

# Notes on Geometry

With 99 Figures

Springer-Verlag
Berlin Heidelberg New York 1983

Professor Elmer G. Rees
University of Edinburgh, Department of Mathematics,
James Clerk Maxwell Building
The King's Buildings, Mayfield Road
Edinburgh, EH9 3JZ, Scotland

ISBN 3-540-12053-X Springer-Verlag Berlin Heidelberg New York
ISBN 0-387-12053-X Springer-Verlag New York Heidelberg Berlin

Library of Congress Cataloging in Publication Data. Main entry under Title: Rees, Elmer G.
Notes on Geometry. (Universitext) Bibliography: p. Includes index.
1. Geometry. I. Title. QA445.R43 1983 516 82-19299



Printing and Binding: fotokop Wilhelm Weihert KG, Darmstadt
2141/3140-543210

# Acknowledgements

I am very grateful to the very many people who have given me encouragement, help and criticism both whilst giving the course and whilst writing these notes. I would like to single out

Christopher Zeeman who, in the early days at Warwick, inspired my interest in geometry of this kind,

Michael Atiyah and Graeme Segal who encouraged me to give this course at Oxford,

Graeme Segal who made numerous suggestions and after my initial attempts gave a much better course than mine. He was very willing for me to incorporate his improvements into these notes. The approach to hyperbolic trigonometry is entirely his,

Wilson Sutherland who saved me from error many times, and finally

Pam Armstrong and David Stewart-Robinson who prepared the text and diagrams.

# Preface

In recent years, geometry has played a lesser role in undergraduate courses than it has ever done. Nevertheless, it still plays a leading role in mathematics at a higher level. Its central role in the history of mathematics has never been disputed. It is important, therefore, to introduce some geometry into university syllabuses. There are several ways of doing this, it can be incorporated into existing courses that are primarily devoted to other topics, it can be taught at a first year level or it can be taught in higher level courses devoted to differential geometry or to more classical topics.

These notes are intended to fill a rather obvious gap in the literature. It treats the classical topics of Euclidean, projective and hyperbolic geometry but uses the material commonly taught to undergraduates: linear algebra, group theory, metric spaces and complex analysis. The notes are based on a course whose aim was two fold, firstly, to introduce the students to some geometry and secondly to deepen their understanding of topics that they have already met. What is required from the earlier material is a familiarity with the main ideas, specific topics that are used are usually redone.

The style of the course was informal and I hope some of the associated good aspects have survived into this version. In line with this, I have taken a concrete viewpoint rather than an axiomatic one. The view that I take is that mathematical objects exist and should be studied, they are not arbitrarily defined as the axiomatic approach might suggest. This is the view of the vast majority of mathematicians in their own work and it is a pity that this does not come across in more undergraduate courses.

There are a large number of exercises throughout the notes, many of these are very staightforward and are meant to test the reader's understanding. Problems, some of them of interest in their own right are given at the end of the three parts. Some are straightforward and some are more like small projects. The more difficult ones are marked with an asterisk.

# Contents

# Introduction

In Euclidean geometry, two triangles are **congruent** if one of them can be moved rigidly onto the other. Definitions such as that of congruence, which tell us when two objects should be regarded as being the same, are basic in geometry and are often used to characterize a particular geometry. Two sets A, B are defined to be equivalent if there is an 'allowed transformation' f such that $fA = B$. For Euclidean geometry the allowed transformations are the **rigid motions**. In his Erlanger programme of 1872, Felix Klein formulated the principle that a geometry is defined by its allowed transformations. The force of this principle is to make a close connection between geometry and group theory.

If S is a set (an example to bear in mind is the Euclidean plane $\mathbf{R}^2$), consider the group Bij(S) consisting of all bijections $f: S \to S$. (If S is a finite set with n elements this is the (familiar) symmetric group $S_n$.) To impose a geometry on S is to consider a subgroup G of Bij(S); two subsets A, B being equivalent for the geometry if there is an $f \in G$ such that $fA = B$. For Euclidean geometry, S is $\mathbf{R}^2$ and G is the group of all rigid motions. Klein's Erlanger programme not only says that the geometry on S and the subgroup G determine each other but that they are, as a matter of definition, one and the same thing. To obtain a worthwhile geometry, the subgroup G has to be chosen with some care after considerable experience. Usually the set S has some structure and the group G preserves this structure, examples are

i) S may be a topological space and the elements of G are homeomorphisms of S.
ii) S may have certain subsets (for example, lines) that are mapped to each other by the elements of G.

There are many other types of examples; in these notes we study the three 'classical' geometries, Euclidean, projective and hyperbolic, but the approach is guided by Klein's Erlanger programme.

# Part I

# Euclidean Geometry

We start by studying the linear groups. These are probably already familiar to the reader. They play an important role in the study of geometry.

**The Linear Groups**

The ring $M(n,\mathbf{R})$ of all $n \times n$ matrices over the field $\mathbf{R}$ of real numbers has the **general linear group** $GL(n,\mathbf{R})$ as its group of units, that is, $GL(n,\mathbf{R})$ consists of all the invertible real $n \times n$ matrices. We will often identify $M(n,\mathbf{R})$ with the space of all linear transformations $T: \mathbf{R}^n \to \mathbf{R}^n$. Note that $M(n,\mathbf{R})$ is a real vector space of dimension $n^2$, and so can be regarded as the metric space $\mathbf{R}^{n^2}$. The determinant defines a continuous map

$$\det: M(n,\mathbf{R}) \to \mathbf{R}$$

(continuous because it is given by a polynomial in the coefficients of a matrix), and $GL(n,\mathbf{R})$ is $\det^{-1}(\mathbf{R} \setminus \{0\})$, so as $\mathbf{R} \setminus \{0\}$ is an open subset of $\mathbf{R}$ we see that

$$GL(n,\mathbf{R}) \text{ is an open subset of } M(n,\mathbf{R}) = \mathbf{R}^{n^2}.$$

The determinant is multiplicative and so defines a homomorphism of groups

$$\det: GL(n,\mathbf{R}) \to \mathbf{R} \setminus \{0\}.$$

Its kernel is the **special linear group** $SL(n,\mathbf{R})$ consisting of matrices with determinant 1. The subset $SL(n,\mathbf{R})$ is closed in $GL(n,\mathbf{R})$ and has dimension $n^2 - 1$ (but is hard to visualize – try to do so for $n = 2$).

Euclidean space $\mathbf{R}^n$ will always be considered with an **inner product** $x.y$ defined on it, this satisfies

i) $(x+y).z = x.z + y.z$ for all $x,y,z \in \mathbf{R}^n$,
ii) $(\lambda x).y = \lambda(x.y)$ for all $x,y \in \mathbf{R}^n$, $\lambda \in \mathbf{R}$,
iii) $x.y = y.x$ for all $x, y \in \mathbf{R}^n$, and
iv) $x.x = 0 \Leftrightarrow x = 0$.

The inner product defines a **norm** $\| \, \|$ on $\mathbf{R}^n$ by $\|x\|^2 = x.x$ and a **metric** d by $d(x,y) = \|x-y\|$. Note that $d(x+a,y+a) = d(x,y)$ so that distance is translation invariant. From the viewpoint of Euclidean geometry the most important transformations are those that preserve distances. We will now study such linear transformations.

A linear transformation $T: \mathbf{R}^n \to \mathbf{R}^n$ is called **orthogonal** if $Tx.Ty = x.y$ for all $x, y \in \mathbf{R}^n$.

A basis $\{e_1, e_2, \ldots, e_n\}$ for $\mathbf{R}^n$ is **orthogonal** if

$$e_i.e_j = 0 \text{ if } i \neq j$$
$$= 1 \text{ if } i = j.$$

If we write $x = \sum_1^n x_i e_i$ and $y = \sum_1^n y_i e_i$, then

$$x.y = \underline{x}^t \underline{y},$$

where on the right hand side $\underline{x}$, $\underline{y}$ denote the column vectors with entries $x_i$, $y_i$ respectively. If T is an orthogonal transformation and A is the matrix of T with respect to an orthonormal basis, an easy calculation shows that

$$\underline{x}^t A^t A \underline{y} = \underline{x}^t \underline{y}$$

and by choosing various suitable x, y one sees that $A^tA = I$. Check this by using the following result.

**Exercise** Note that $a_{ij} = e_i{}^t A e_j$ and hence show that if $x^tAy = x^tBy$ for all $x,y \in \mathbf{R}^n$ then $A = B$.

The matrix of an orthogonal transformation with respect to an orthogonal basis is therefore orthogonal in the usual sense for matrices. Note that a little care is needed in handling orthogonal matrices because a matrix is orthogonal if and only if its columns (or rows) form an **orthonormal** set of vectors.

If X is a metric space, a map $f: X \to X$ is an **isometry** if it is onto and distance preserving, that is, $d(fx,fy) = d(x,y)$ for all $x,y \in X$.

**Exercise** i) Show that a distance preserving map is one to one.

ii) Show that a distance preserving map $f: X \to X$ is not necessarily onto by considering the map $f: \mathbf{R}_+ \to \mathbf{R}_+$ defined by $f(x) = x+1$.

**Exercise** If X is a metric space, verify that the set of all isometries $f: X \to X$ forms a group under composition.

The properties of the isometries of a metric space X are intimately connected with the properties of X itself. The importance of orthogonal transformations in Euclidean geometry arises because they are isometries of $\mathbf{R}^n$, moreover, apart from translations, they are in a sense all the isometries.

**Lemma** i) If $T: \mathbf{R}^n \to \mathbf{R}^n$ is a linear isometry then T is orthogonal.

ii) If $T: \mathbf{R}^n \to \mathbf{R}^n$ is linear and norm-preserving then T is orthogonal.

**Proof** Notice that any linear isometry is norm preserving because any such isometry satisfies

$$T(x-y).T(x-y) = (Tx-Ty).(Tx-Ty) = d(Tx,Ty)^2 = d(x,y)^2 = (x-y).(x-y)$$

so by putting $y = 0$ one gets

$$\|Tx\|^2 = \|x\|^2.$$

Hence it suffices to prove ii). The map T preserves the norm of $x-y$ so

$$T(x-y).T(x-y) = (x-y).(x-y).$$

Expanding these expressions using linearity gives

$$Tx.Tx - 2Tx.Ty + Ty.Ty = x.x - 2x.y + y.y.$$

But T also preserves the norms of x and y, so

$$Tx.Ty = x.y \text{ for all } x, y \in \mathbf{R}^n,$$

and so T is orthogonal.

Of course there are isometries that are not linear, for example, the translations $T_a: \mathbf{R}^n \to \mathbf{R}^n$ defined by $T_a(x) = a + x$ are not linear unless $a = 0$. Later we will show that any isometry $f: \mathbf{R}^n \to \mathbf{R}^n$ such that $f0 = 0$ is linear.

The set of all orthogonal $n \times n$ matrices form the **orthogonal group**

$$O(n) = \{A \in GL(n,\mathbf{R}) \mid A^tA = I\}.$$

If A is orthogonal then $\det A = \pm 1$ because $\det A^t = \det A$ and so

$$1 = \det I = \det(A^tA) = \det A^t.\det A = (\det A)^2.$$

The group O(n) has a normal subgroup, the **special orthogonal group** $SO(n) = O(n) \cap SL(n,\mathbf{R})$ consisting of the orthogonal matrices whose determinant is $+1$. This subgroup of O(n) has index 2.

**Examples** $O(1) = \{\pm 1\}, SO(1) = \{+1\}$.

The group O(2) consists of $2 \times 2$ matrices $\begin{bmatrix} a & b \\ c & d \end{bmatrix}$ whose columns are orthonormal. An elementary calculation shows that there is a $\theta$ such that

$$\begin{bmatrix} a \\ c \end{bmatrix} = \begin{bmatrix} \cos\theta \\ \sin\theta \end{bmatrix} \text{ and } \begin{bmatrix} b \\ d \end{bmatrix} = \pm \begin{bmatrix} -\sin\theta \\ \cos\theta \end{bmatrix},$$

SO(2) consists of the matrices $\begin{bmatrix}\cos\theta & -\sin\theta\\ \sin\theta & \cos\theta\end{bmatrix}$. This matrix represents a rotation through the angle $\theta$ about the origin. As SO(2) has index 2 in O(2), it has two cosets, one is SO(2) itself and the other is SO(2). $\begin{bmatrix}1 & 0\\ 0 & -1\end{bmatrix}$, which consists of the matrices $\begin{bmatrix}\cos\theta & \sin\theta\\ \sin\theta & -\cos\theta\end{bmatrix}$. This matrix represents a reflection $R_\ell$ in the line $\ell$ which makes the angle $\theta/2$ with the x-axis.

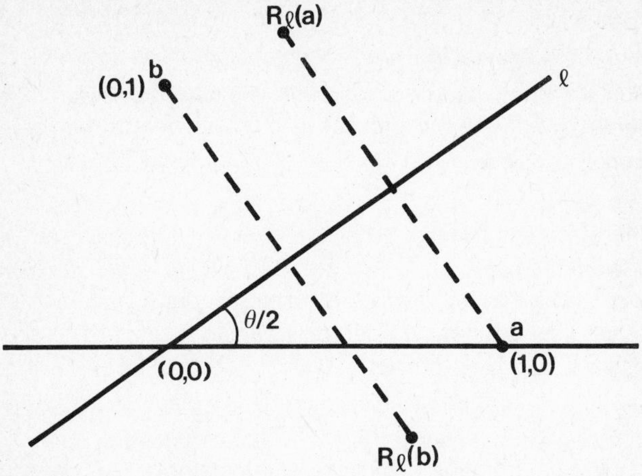


**The relationship between O(n) and GL(n,R)**

A matrix in GL(n,**R**) has independent columns and a matrix in O(n) has orthonormal columns. The **Gram-Schmidt process** transforms an independent set of vectors into an orthonormal set, so it can be used to define a mapping GL(n,**R**) $\to$ O(n). To make this precise it is convenient to introduce the group $T_+$ (n) consisting of the set of **upper triangular** $n \times n$ matrices whose diagonal entries are positive (T for triangular, + for positive).

**Proposition** $T_+$ (n) is a subgroup of GL(n,**R**).

**Proof** If $A \in T_+(n)$, then $\det A = a_{11}\, a_{22} \ldots a_{nn} > 0$. Hence $T_+(n)$ is a subset of GL(n,**R**). The matrix A lies in $T_+$(n) if and only if

$$a_{ij} = 0 \text{ for } i > j$$
$$\text{and} \qquad a_{ii} > 0.$$

If A, B $\in T_+$ (n) then

$$(AB)_{ij} = \sum_{k=1}^{n} a_{ik} b_{kj}.$$

If $i > j$, then either $i > k$ or $k \geq i > j$; in the first case $a_{ik} = 0$ and in the second $b_{kj} = 0$, so $(AB)_{ij} = 0$ in both cases. If $i = j$, then $(AB)_{ii} = a_{ii}b_{ii} > 0$. So AB $\in$ T (n) if A, B $\in$ T (n).

It remains to check that if $A \in T_+(n)$ then so is $A^{-1}$. Suppose $A \in T_+$ (n) and $AB = BA = I$. We show that $b_{ij} = 0$ for $i > j$ by downward induction on i. First consider $i = n$, then

$$0 = a_{nn} b_{nj} \text{ for } j < n, \text{ so } b_{nj} = 0.$$

Suppose that $b_{kj} = 0$ for all $j < k$ if $k > i$, then

$$0 = a_{ii} b_{ij} \text{ for } i > j, \text{ so } b_{ij} = 0.$$

Given that $b_{ij} = 0$ for $i > j$, one gets that

$$a_{ii}b_{ii} = 1 \text{ for all } i.$$

Hence as $a_{ii} > 0$ one sees that $b_{ii} > 0$. Hence $B \in T_+(n)$.

(The reader may prefer to go through this proof explicitly in the case $n = 2$.)

**Theorem 1** For a given $A \in GL(n,\mathbf{R})$, there are unique matrices $B \in O(n)$, $C \in T_+(n)$ such that $A = BC$.

**Proof** As suggested above, we use the Gram-Schmidt process to construct B from A and then observe that they are related by $A = BC$ with $C \in T_+(n)$. In detail: let $a_1, a_2, \ldots, a_n$ be the columns of A. The first stage of the Gram-Schmidt process is to find an orthogonal set $f_1, f_2, \ldots f_n$. This is constructed by induction as follows.

$$f_1 = a_1,$$
$$f_k = a_k - \sum_{i=1}^{k-1} \{(a_k.f_i)/(f_i.f_i)\}f_i.$$

If F is the matrix with columns $f_1, f_2, \ldots, f_n$ then $F = AT_1$, where $T_1$ is in $T_+(n)$, and in fact $T_1$ has ones along the diagonal. Note that the matrix F is obtained from A by a sequence of elementary column operations, each new column involving only earlier columns. The second stage of the Gram-Schmidt process is to normalise the set $f_1, f_2, \ldots, f_n$, that is, let $b_i = f_i / \|f_i\|$. If B is the matrix whose columns are $b_1, b_2, \ldots, b_n$ then $B = FT_2$ where $T_2$ is a diagonal matrix with positive entries $(1 / \|f_i\|)$ on the diagonal, hence $T_2 \in T_+(n)$. If $C = (T_1T_2)^{-1}$ we have $A = BC$ with $B \in O(n)$ and $C \in T_+(n)$. Moreover, it is clear from the formulae that the matrices B, C depend continuously on the original matrix A.

It remains to check the uniqueness of this decomposition. Suppose A has two such decompositions, $B_1C_1$ and $B_2C_2$ say. Then $D = B_2^{-1}B_1 = C_2C_1^{-1}$ is in $O(n) \cap T_+(n)$. But we will show that $O(n) \cap T_+(n) = \{I\}$ and so the decomposition is unique. Let $D \in O(n) \cap T_+(n)$ then $D^t = D^{-1}$ and as $T_+(n)$ is a subgroup we have $D^t \in T_+(n)$. But $D^t$ is lower triangular so D must be diagonal, and therefore $D = D^t$, so using orthogonality, $D^2 = I$. So D has diagonal entries $\pm 1$. As $D \in T_+(n)$ it has positive entries on the diagonal therefore $D = I$ as required.

**Corollary** $GL(n,\mathbf{R})$ is homeomorphic to $O(n) \times T_+(n)$.

**Proof** The homeomorphisms are constructed as follows: $A \in GL(n,\mathbf{R})$ is mapped to $(B,C)$ and $(B,C) \in O(n) \times T_+(n)$ is mapped to BC. These are clearly mutual inverses. The map $(B,C) \to BC$ is continuous because matrix multiplication is continuous – the entries of BC are polynomials in the entries of B and C. The map $A \to (B,C)$ is also continuous for a similar reason.

**Note** $T_+(n)$ is homeomorphic to $\mathbf{R}^{n(n+1)/2}$. A matrix in $T_+(n)$ has $n(n-1)/2$ entries off the diagonal and each of these can be an arbitrary element in $\mathbf{R}$. There are n entries on the diagonal and each of these is an arbitrary element of $\mathbf{R}_+$, so that $T_+(n) \cong \mathbf{R}^{n(n+1)/2} \times (\mathbf{R}_+)^n$. But $\mathbf{R}_+$ is homeomorphic to $\mathbf{R}$ (under log and exp as inverse homeomorphisms).

So $GL(n,\mathbf{R})$ is homeomorphic to $O(n) \times \mathbf{R}^{n(n+1)/2}$.

**Exercise** The space $SL(n,\mathbf{R})$ is homeomorphic to $SO(n) \times \mathbf{R}^{(n^2+n-2)/2}$.

**Examples** The space $GL(1,\mathbf{R})$ is $\mathbf{R} \setminus \{0\}$, $O(1)$ is $\{\pm 1\}$ and so one can see directly that $GL(1,\mathbf{R})$ is homeomorphic to $O(1) \times \mathbf{R}$.

The group $O(2)$ is the union of $SO(2)$ and another coset of $SO(2)$ but $SO(2)$ is homeomorphic to the **circle** $S^1 = \{z \in \mathbf{C} \mid \|z\| = 1\}$, so $O(2)$ is homeomorphic to the union of two disjoint copies of $S^1$ and $GL(2,\mathbf{R})$ is homeomorphic to the union of two disjoint copies of $S^1 \times \mathbf{R}^3$.

**Exercise** Show that the group $GL(n,\mathbf{R})$ for $n > 1$ is not the direct product of its subgroups $O(n)$ and $T_+(n)$.

**Exercise** Show that GL(2,**R**) has many subgroups of order three but that $O(2) \times T_+(2)$ has only one such subgroup. Deduce that there is no isomorphism between GL(2,**R**) and $O(2) \times T_+(2)$. [It is true that GL(n,**R**) is not isomorphic to $O(n) \times T_+(n)$ for any $n \geq 2$, but the proof is more difficult for $n \geq 3$. For $n \geq 3$, show that the centre Z of G = GL(n,**R**) consists of the scalar matrices and that G/Z has no proper normal subgroups for n odd and only one such for n even. The group $O(n) \times T_+(n)$ modulo its centre has several proper normal subgroups.]

**Affine Subspaces and Affine Independence**

It is often necessary to consider lines, planes, etc. that do not pass through the origin. Linear subspaces always contain the origin but their cosets (in the additive group) do not and they are called affine subspaces. However the affine subspaces have an intrinsic definition.

A subset A of $\mathbf{R}^n$ is an **affine subspace** if $\lambda a + \mu b \in A$ for all $a, b \in A$ and all $\lambda, \mu \in \mathbf{R}$ such that $\lambda + \mu = 1$. A straightforward induction shows that the following is an equivalent condition:

$$\sum_{i=1}^{k} \lambda_i a_i \in A \text{ for all } a_i \in A \text{ and } \sum_{i=1}^{k} \lambda_i = 1.$$

If $V \subset \mathbf{R}^n$ is a linear subspace then it is easy to check that the set $V + x = \{v+x \mid v \in V\}$ is an affine subspace of $\mathbf{R}^n$ for any (fixed) $x \in \mathbf{R}^n$. Every affine subspace A is of this form, because if $a \in A$ and $V = A - a = \{x-a \mid x \in A\}$ then V is a linear subspace of $\mathbf{R}^n$. Let $\lambda \in \mathbf{R}$ and $x - a \in V$ then to check that $\lambda(x-a) \in V$ we must check that $\lambda(x-a) + a \in A$ but

$$\lambda(x-a) + a = \lambda x + (1-\lambda)a$$

and $x, a \in A$. If $x-a, y-a \in V$ then $(x-a) + (y-a) \in V$ because

$$(x-a) + (y-a) + a = x + y - a$$

which is a linear combination of elements of A, the sum of the coefficients being $1 + 1 - 1 = 1$ so $x + y - a \in A$.

**Exercise** If $a, b \in A$ and A is an affine subspace, show that $A - a = A - b$.

If A is an affine subspace, its **dimension** is the dimension of the linear subspace $A - a$ of $\mathbf{R}^n$.

If $X \subset \mathbf{R}^n$ is any subset, its **affine span** Aff(X) is defined as

$$\mathrm{Aff}(x) = \{\sum_{i=1}^{k} \lambda_i x_i \mid x_i \in X, \sum_{i=1}^{k} \lambda_i = 1\}.$$

It is easy to check that Aff(x) is an affine subspace of $\mathbf{R}^n$ and that it is the smallest affine subspace containing X.

A set $X = \{x_0, x_1, \ldots, x_r\}$ is **affinely independent** if $\sum_{i=0}^{r} \lambda_i x_i = 0$ holds with $\sum_{i=0}^{r} \lambda_i = 0$ only if $\lambda_0 = \lambda_1 = \ldots = \lambda_r = 0$. It is easily checked that $\{x_0, x_1, \ldots, x_r\}$ is affinely independent if and only if $\{x_1-x_0, x_2-x_0, \ldots, x_r-x_0\}$ is a linearly independent set.

If $X = \{x_0, x_1, \ldots, x_r\}$ is affinely independent then Aff(X) has dimension r and X is called an **affine basis** for Aff(X). Note that an affine basis for an r-dimensional affine subspace has $r + 1$ elements. If $\{e_1, e_2, \ldots, e_r\}$ is a basis of a linear subspace V then an affine basis for V is $\{0, e_1, e_2, \ldots, e_r\}$.

An affine subspace H of $\mathbf{R}^n$ whose dimension is $n - 1$ is called a **hyperplane**. If H is a linear hyperplane of $\mathbf{R}^n$, then there is a non-zero $x \in \mathbf{R}^n$ such that $H = \{x\}^\perp$. This is because one can choose an orthonormal basis for H and extend it (by a vector x) to an orthonormal basis for $\mathbf{R}^n$; it is then easy to check that $H = \{x\}^\perp$. Hyperplanes arise as the perpendicular bisectors of line segments.

**Lemma** If $a, b \in \mathbf{R}^n$ with $a \neq b$, then $B = \{x \mid d(x,a) = d(x,b)\}$ is a hyperplane in $\mathbf{R}^n$.

**Proof** It is clear that $(a+b)/2 \in B$ so we need to show that $H = B - (a+b)/2$ is an (n–1) dimensional linear subspace. If $c = (a-b)/2$, it is easily checked using the translation invariance of distance that H is the set $\{x \mid d(x,c) = d(x,-c)\}$. If $c, e_2, e_3, \ldots, e_n$ is an orthogonal basis for $\mathbf{R}^n$, then

$e_2, e_3, \ldots, e_n$ is a basis for H.

If H is any hyperplane in $\mathbf{R}^n$ and $x \in \mathbf{R}^n$, then x can be written uniquely in the form y + z where $y \in H$ and $z \perp H$.

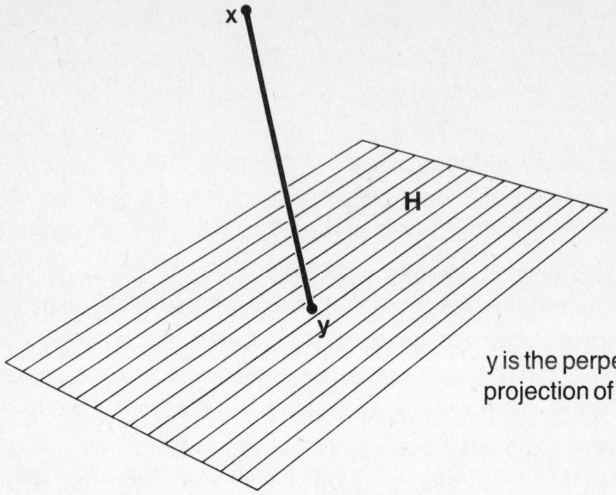


y is the perpendicular projection of x onto H.

More algebraically, let $a \in H$, then $H - a$ is a linear hyperplane, so $H - a = \{b\}^\perp$ for some b. There is a unique expression

$$x - a = \lambda b + c \text{ where } c \in H - a.$$

Let $y = c + a$, $z = \lambda b$, then $y \in H$ and $z \perp H$.

It remains to check the uniqueness. Suppose

$$y_1 + z_1 = y_2 + z_2 \quad \text{with} \quad y_1, y_2 \in H, z_1, z_2 \in H^\perp$$

Then $z_2 - z_1 = y_1 - y_2 \in H - y_2$ and $z_2 - z_1 \in H - y_2$, hence $z_1 = z_2$ and so $y_1 = y_2$.

If $A \subset \mathbf{R}^n$, $B \subset \mathbf{R}^m$ are both affine subspaces, a map $f: A \to B$ is an **affine map** if

$$f(\lambda a + \mu b) = \lambda f(a) + \mu f(b) \text{ for } a, b \in A \text{ and } \lambda + \mu = 1.$$

An affine map is therefore one that takes straight lines to straight lines because the straight line through the points a, b is the set $\{\lambda a + \mu b | \lambda, \mu \in \mathbf{R}, \lambda + \mu = 1\}$. If $a \in A$, $b \in B$ then A–a and B–b are linear spaces; if $L: A{-}a \to B{-}b$ is a linear map then the map $A_L: A \to B$ defined by

$$A_L(x) = L(x-a) + b$$

is an affine map. When checking this note carefully that L is only defined on $A - a$. All affine maps arise in this manner as we now show.

**Lemma** If $f: A \to B$ is an affine map then the map $L_f: A{-}a \to B{-}f(a)$ defined by

$$L_f(x) = f(x+a) - f(a)$$

is a linear map. The map f is obtained from $L_f$ by the previous construction.

**Proof** We need to check that $L_f(x+y) = L_f(x) + L_f(y)$ and that $L_f(\lambda x) = \lambda L_f(x)$. To check the first we note that $x + a$, $y + a$, $a \in A$ and that $x + y + a = (x+a) + (y+a) - a$ is a combination of them, the sum of whose coefficients is 1. So

$$\begin{aligned} L_f(x+y) &= f(x+y+a) - f(a) \\ &= f(x+a) + f(y+a) - f(a) - f(a) \\ &= L_f(x) + L_f(y). \end{aligned}$$

$$
\begin{aligned}
L_f(\lambda x) &= f(\lambda x+a)-f(a)\\
&= f(\lambda(x+a)+(1-\lambda)a)-f(a)\\
&= \lambda f(x+a) + (1-\lambda)f(a)-f(a)\\
&= \lambda L_f(x).
\end{aligned}
$$

If $L = L_f$, it is easy to check that if one takes $b = f(a)$ then $f = A_L$.

An important special case is the following.

**Corollary** If $f: \mathbf{R}^n \to \mathbf{R}^n$ is an affine map then there exists $a \in \mathbf{R}^n$ such that the map $L: \mathbf{R}^n \to \mathbf{R}^n$ defined by $L(x) = f(x) - a$ is linear, so $f(x) = L(x) + a$.

**Isometries of $\mathbf{R}^n$**

We have already seen that translations $T_a: \mathbf{R}^n \to \mathbf{R}^n$ defined by $T_a(x) = x + a$ and orthogonal transformations $T: \mathbf{R}^n \to \mathbf{R}^n$ which are linear and satisfy $Tx.Ty = x.y$ are both examples of isometries of $\mathbf{R}^n$. We will show that all isometries are combinations of these two basic kinds. In fact isometries are affine maps. The first step is

**Theorem 2** An isometry $f: \mathbf{R}^n \to \mathbf{R}^n$ is uniquely determined by the images $fa_0, fa_1, \ldots fa_n$ of a set $a_0, a_1, \ldots a_n$ of $(n+1)$ (affinely) independent points.

**Proof** Let f, g be isometries with $fa_i = ga_i$ for $0 \leqslant i \leqslant n$. Then $g^{-1}f$ is an isometry with $g^{-1}fa_i = a_i$. Let T be the translation defined by $Tx = x - a_0$ and let $b_i = T(a_i)$ for $0 \leqslant i \leqslant n$. Clearly, $b_0 = 0$, and the set $\{b_1, b_2, \ldots, b_n\}$ forms a basis for $\mathbf{R}^n$. We will show that $h = Tg^{-1}fT^{-1}$ is the identity, and this shows that $f = g$ as required.

Clearly $hb_i = b_i$ for $0 \leqslant i \leqslant n$, so if $y = hx$ we have that $d(x,0) = d(y,0)$ and $d(x,b_i) = d(y,b_i)$ for $1 \leqslant i \leqslant n$ because h is an isometry. Hence $x.x = y.y$ and $(x-b_i).(x-b_i) = (y-b_i).(y-b_i)$ for $1 \leqslant i \leqslant n$. By expanding these last n equations and manipulating one gets that $x.b_i = y.b_i$ for $1 \leqslant i \leqslant n$. As $b_1, b_2, \ldots, b_n$ is a basis, one has $x.z = y.z$ for every $z \in \mathbf{R}^n$, hence $x = y$, proving that h is the identity.

This proof shows that a point in $\mathbf{R}^n$ is uniquely determined by its distances from $n + 1$ independent points. Note that, in general, a point x is not uniquely determined by its distances from n independent points.

**Theorem 3** If $\{a_0, a_1, \ldots a_n\}$ and $\{b_0, b_1, \ldots b_n\}$ are two sets of $(n+1)$ independent points in $\mathbf{R}^n$ with $d(a_i,a_j) = d(b_i,b_j)$ for $0 \leqslant i, j \leqslant n$ then there is an isometry $f: \mathbf{R}^n \to \mathbf{R}^n$ with $fa_i = fb_i$ for $0 \leqslant i \leqslant n$.

**Proof** Using translations we can clearly assume that $a_0 = b_0 = 0$. Then $\{a_1, a_2, \ldots a_n\}$ and $\{b_1, b_2, \ldots b_n\}$ are bases for $\mathbf{R}^n$, and it is easy to see that the hypotheses imply that $a_i.a_j = b_i.b_j$ for all i, j. Let g be the unique (non-singular) linear transformation such that $ga_i = gb_i$ for $1 \leqslant i \leqslant n$. Let $x - y = \Sigma\, \lambda_i a_i$, then $gx - gy = g(x-y) = \Sigma\, \lambda_i b_i$ by the linearity of g. So

$$d(gx,gy)^2 = \Sigma\, \lambda_i\lambda_j b_i.b_j = \Sigma\, \lambda_i\lambda_j a_i.a_j = d(x,y)^2.$$

Hence g is a linear isometry. The required f is the composition of g with a translation, so it is affine.

We have already proved on page 4 that every linear isometry is orthogonal. Theorems 2 and 3 therefore combine to show that if f is an isometry of $\mathbf{R}^n$ then $f(x) = Ax + a$ where $A \in O(n)$ and $a \in R^n$ so that every isometry is the composition of an orthogonal transformation and a translation. Hence every isometry is affine.

**Exercise** If $X \subset \mathbf{R}^n$ is any subset and $g: X \to \mathbf{R}^n$ is an isometric map, show that there is an isometry $f: \mathbf{R}^n \to \mathbf{R}^n$ such that $f|X$ is g. If the affine subspace defined by X has dimension $n - r$, prove that the set of such isometries forms a coset of $O(r)$.

We will now show how every isometry can be written as a product of reflections. This gives an alternative approach to understanding isometries and yields independent proofs of some of our previous results.

**Definition** If H is a hyperplane in $\mathbf{R}^n$, **reflection in H** is the isometry $R_H$ of $\mathbf{R}^n$ defined by

$$R_H(x) = y - z$$

where $x = y + z$ with $y \in H$ and $z \perp (H-y)$ using the decomposition given on page 8.

Note that $R_H{}^2$ is the identity and that $R_H$ leaves every point of H fixed. If H is the perpendicular bisector of ab, $R_H$ interchanges a and b.

**Example** In $\mathbf{R}^3$, regard H as a two-sided mirror, then $R_H(x)$ is the mirror image of x.

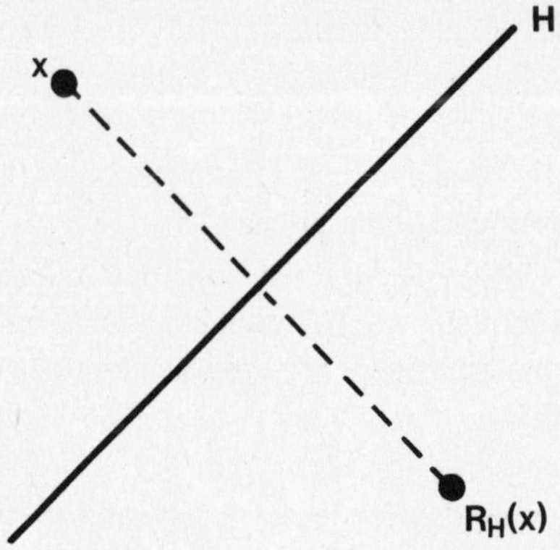


**Exercise** If $0 \in H$, that is if H is a linear hyperplane, show that $R_H$ is orthogonal. If a is a unit vector perpendicular to H, show that $R_H(x) = x - 2(x.a)a$.

**Theorem 4** Any isometry $f: \mathbf{R}^n \to \mathbf{R}^n$ that is the identity on an affine (n–r)-dimensional subspace A (that is, $fa = a$ for each $a \in A$) can be expressed as the product of at most r reflections in hyperplanes that contain A. Any isometry can be expressed as the product of at most (n+1) reflections.

**Note** The last sentence can be regarded as a special case of the first if one makes the (usual) convention that the empty set has dimension –1.

**Proof** Choose (n–r+1) independent points $a_0, a_1, \ldots, a_{n-r}$ in A and extend them to a set $a_0, a_1, \ldots, a_n$ of (n+1) independent points in $\mathbf{R}^n$. Let $b_i = fa_i$, so $b_i = a_i$ for $0 \leqslant i \leqslant n-r$. As f is an isometry, $d(a_i,a_j) = d(b_i,b_j)$ so if H is the perpendicular bisector of $a_{n-r+1}b_{n-r+1}$ it is clear that $a_i \in H$ for $0 \leqslant i \leqslant n-r$. The idea now is to consider $R_Hf$, this is the identity on an (n–r+1) dimensional affine subspace and so one can use induction on r to give the required result. In detail: $R_Hf = R_{H_1} \ldots R_{H_s}$ where $H_1 \ldots H_s$ are s hyperplanes ($s \leqslant r-1$) containing A and $a_{n-r+1}$, then $f = R_H R_{H_1} \ldots R_{H_s}$ is a product of at most r reflections in hyperplanes containing A. To prove the last sentence of the theorem, let H be the hyperplane bisecting af(a) for some $a \in \mathbf{R}^n$. Then $R_Hf$ fixes a 0-dimensional affine subspace, and so is the product of at most n reflections. So f is the product of at most n + 1 reflections.

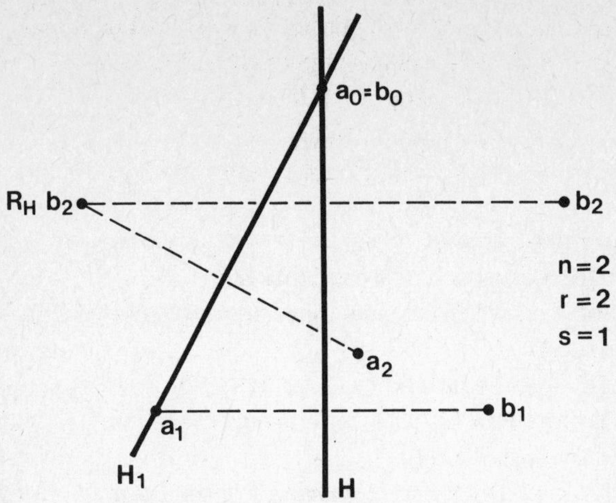


**Corollary 1** If $f: \mathbf{R}^n \to \mathbf{R}^n$ is an isometry with $f(0) = 0$ then f is orthogonal.

**Proof** The theorem proves that f is the product of at most n reflections in hyperplanes through 0. But such reflections are orthogonal, and hence so is their product.

**Corollary 2** Any isometry $f: \mathbf{R}^n \to \mathbf{R}^n$ is the composition of an orthogonal transformation and a translation. In fact $T_{-f0}.f$ and $f.T_{f^{-1}0}$ are orthogonal.

**Proof** $T_{-f0}.f$ and $f.T_{f^{-1}0}$ both fix 0 so the result follows from Corollary 1.

We will denote the **group of all isometries of $\mathbf{R}^n$** by $\mathbf{I}(\mathbf{R}^n)$.

**Exercise** If $a \in \mathbf{R}^n$, the translation $T_a$ is defined by $T_a(x) = x + a$. Show that the set of all translations forms a normal subgroup of $\mathbf{I}(\mathbf{R}^n)$ which is isomorphic to the additive group $\mathbf{R}^n$ and whose quotient is isomorphic to $O(n)$.

A metric d can be defined on $\mathbf{I}(\mathbf{R}^n)$ as follows:
Choose a set of $(n+1)$ independent points $a_0, a_1, \ldots, a_n$ in $\mathbf{R}^n$ and let

$$d(f,g) = \max_{0 \leq i \leq n} d(fa_i, ga_i) \qquad \text{for } f,g \in \mathbf{I}(\mathbf{R}^n).$$

By definition, $d(f,g) = 0$ if and only if $fa_i = ga_i$ for $0 \leq i \leq n$ and by Theorem 2, this holds precisely when $f = g$. The triangle inequality is easy to check. This metric has the further property of left invariance:

$$d(hf,hg) = d(f,g) \qquad \text{for all } f, g, h \in \mathbf{I}(\mathbf{R}^n).$$

**Exercise** On the subset of $\mathbf{I}(\mathbf{R}^n)$ consisting of all the translations, show that d gives the usual metric on $\mathbf{R}^n$.

Different sets of independent points give different metrics on $\mathbf{I}(\mathbf{R}^n)$. These metrics give rise to the same topology on $\mathbf{I}(\mathbf{R}^n)$:

If $a = \{a_o, a_1, \ldots, a_n\}$ and $b = \{b_o, b_1, \ldots, b_n\}$ give rise to metrics $d_a$ and $d_b$ respectively, then $d_a$ and $d_b$ both define the same topology. By symmetry between a and b it is enough to prove that for every $\varepsilon > 0$ there is a $\delta > 0$ such that

$$d_b(f,g) < \delta \text{ implies that } d_a(f,g) < \varepsilon.$$

First write the $a_i$ vectors in terms of the $b_i$ vectors:

$$a_i = \sum_{j=0}^{n} \lambda_{ij} b_j \text{ with } \sum_{j=0}^{n} \lambda_{ij} = 1.$$

Choose M so that $|\lambda_{ij}| \leq M$ for all i, j and now choose $\delta > 0$ so that $(n+1)M\delta < \varepsilon$. If $d_b(f,g) < \delta$ then $\|fb_i - gb_i\| < \delta$ for each i. As f, g are both affine maps one has that

$$fa_i - ga_i = \sum_{i=0}^{n} \lambda_{ij}(fb_i - gb_i).$$

So

$$\|fa_i - ga_i\| \leq \sum_{i=0}^{n} |\lambda_{ij}| \; \|fb_i - gb_i\| < (n+1)M\delta < \varepsilon.$$

The result follows.

Corollary 2 above implies the following

**Theorem 5** The space $\mathbf{I}(\mathbf{R}^n)$ is homeomorphic to $O(n) \times \mathbf{R}^n$.

**Proof** We define the two maps which are mutual inverses and leave the verification that they are both continuous to the reader.

Given $f \in \mathbf{I}(\mathbf{R}^n)$, let $\bar{f}: \mathbf{R}^n \to \mathbf{R}^n$ be defined by $\bar{f}(x) = f(x) - f(0)$, so that $\bar{f} = T_{-f0}.f$. Then $\bar{f}(0) = 0$, so that $\bar{f}$ is orthogonal. The map $\mathbf{I}(\mathbf{R}^n) \to O(n) \times \mathbf{R}^n$ is now defined by $f \to (\bar{f}, f(0))$, and its inverse by $(T,a) \to f$, where $f(x) = T(x) + a = T_aT(x)$.

**Exercise** Prove that the group $\mathbf{I}(\mathbf{R}^n)$ is not isomorphic to the direct product of its subgroups $O(n)$ and $\mathbf{R}^n$ (the subgroup of translations). [Hint: find $f \in O(n)$ and $g \in \mathbf{R}^n$ so that $fg \neq gf$.]

The mapping $f \to \bar{f}: \mathbf{I}(\mathbf{R}^n) \to O(n)$ is useful because it can be used to distinguish between two different types of isometries.

**Definition** An isometry f is called **direct** or **opposite** according as $\det \bar{f} = +1$ or $-1$ where $\bar{f}(x) = f(x) - f(0)$ ($\bar{f}$ is orthogonal, so that one knows $\det \bar{f} = \pm 1$ for any isometry f).

**Care** Sometimes in the literature (especially in physics textbooks) these are called proper and improper motions.

**Lemma** The map $\mathbf{I}(\mathbf{R}^n) \to \{\pm 1\}$ defined by $f \to \det \bar{f}$ is a group homomorphism.

**Proof** It is clearly enough to show that the map $f \to \bar{f}$ is a group homomorphism because det is a multiplicative map. We show that $\overline{fg} = \bar{f}\bar{g}$.

$$\begin{aligned}
\bar{f}(\bar{g}(x)) &= \bar{f}(g(x) - g(0)) \\
&= \bar{f}(g(x)) - \bar{f}(g(0)) \qquad \text{by linearity of } \bar{f} \\
&= f(g(x)) - f(0) - (f(g(0)) - f(0)) \\
&= f(g(x)) - f(g(0)) = \overline{fg}(x).
\end{aligned}$$

**Exercise** It is worthwhile to compare this 'sign' of an isometry with the sign of a permutation (see page 28). Define a monomorphism $i: S_n \to O(n)$ by permuting the axes of $\mathbf{R}^n$. Show that $\det(i(\sigma))$ equals the sign of the permutation $\sigma$. (It is enough to check this for transpositions.)

**Exercise** (A refinement of Theorem 2).

Show that an isometry of $\mathbf{R}^n$ is uniquely determined by the images of n independent points together with its sign.

**Isometries of $\mathbf{R}^2$**

We will now study the isometries of $\mathbf{R}^2$ in some detail. First, we have the examples, we list their properties and the reader should verify them.

**Direct**

1. **Translations** $T_a(x) = x + a$.

The set of all translations forms a normal subgroup of $\mathbf{I}(\mathbf{R}^2)$ and is isomorphic with $\mathbf{R}^2$. The translation $T_a$ is the product of two reflections in lines $\ell$, m both perpendicular to a and distance $\|a\| / 2$ apart. $T_a$ is an element of infinite order in $\mathbf{I}(\mathbf{R}^2)$.

2. **Rotations** $R(a,\alpha)$ is the rotation through angle $\alpha$ about the point a. The rotation $R(0,\alpha)$ has matrix $\begin{bmatrix} \cos\alpha & -\sin\alpha \\ \sin\alpha & \cos\alpha \end{bmatrix}$. Note that $fR(a,\alpha)f^{-1} = R(fa,\alpha)$ for a direct isometry f and that $fR(a,\alpha)f^{-1} = R(fa,-\alpha)$ for an opposite isometry f.

For a fixed a, the set of all rotations $R(a,\alpha)$ forms a subgroup $SO(2)_a$ of $\mathbf{I}(\mathbf{R}^2)$ which is isomorphic to $SO(2)$ ($=SO(2)_0$), and, for different a, each of these subgroups are conjugate. The rotation $R(a,\alpha)$ is the product of two reflections in lines through a with angle $\alpha/2$ between them. $R(a,\alpha)$ has infinite order when $2\pi/\alpha$ is irrational, and in this case it and its powers form a dense subgroup of $SO(2)$. When $2\pi/\alpha$ is rational, $R(a,\alpha)$ has finite order equal to the numerator of $2\pi/\alpha$, and it and its powers form a discrete subgroup of $SO(2)_a$.

**Opposite**

3. **Reflections** $R_\ell$ is reflection in the line $\ell$. The identity and $R_\ell$ form a subgroup of order 2. For any pair of lines $\ell$, m in the plane there is an isometry f such that $f\ell = m$; also $fR_\ell f^{-1} = R_{f\ell}$ for any isometry f. Hence any two subgroups of $\mathbf{I}(\mathbf{R}^2)$ generated by a single reflection are conjugate.

4. **Glides(or glide reflections)** A glide is of the form $G(\ell,a) = R_\ell T_a$ where a is a vector parallel to $\ell$. Note that in this case $T_a R_\ell = R_\ell T_a$.

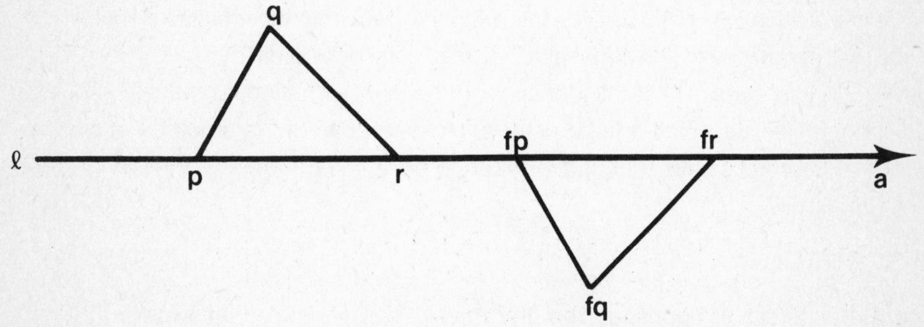


The square of a glide is a translation, $(R_\ell T_a)^2 = T_{2a}$ and so a glide has infinite order.

**Exercise** i) Show that the translations $T_a$, $T_b$ are conjugate in $\mathbf{I}(\mathbf{R}^2)$ if and only if $\|a\| = \|b\|$.

ii) Show that the glides $R_\ell T_a$, $R_m T_b$ are conjugate in $\mathbf{I}(\mathbf{R}^2)$ if and only if $\|a\| = \|b\|$.

Any isometry of $\mathbf{R}^2$ is either the identity or of one of the above four types. To prove this we will need the following result about compositions.

**Lemma** $T_a R_\ell$ is a glide if the line $\ell$ is not perpendicular to a and a reflection if $\ell$ is perpendicular to a.

**Proof**

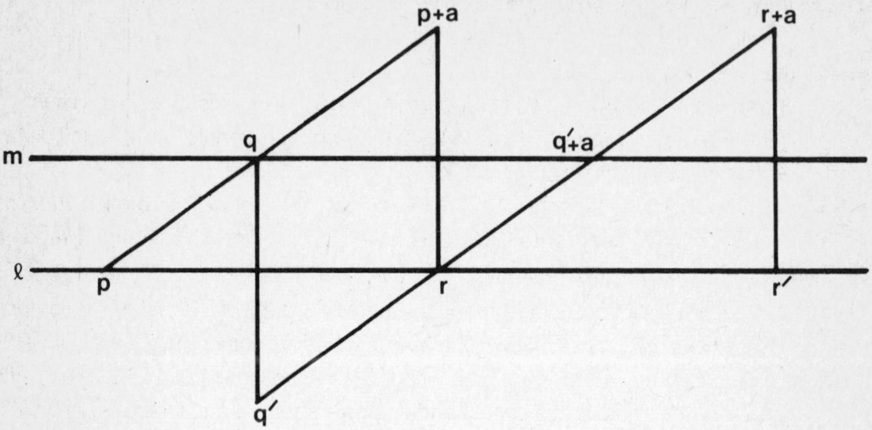


Choose $p \in \ell$, let $q = p + a/2$ and let r be the foot of the perpendicular from $p + a$ to $\ell$. Let m denote the line parallel to $\ell$ through q. We prove that $T_a R_\ell = R_m T_{r-p}$, by checking that these transformations are the same on the three independent points p, q, r (the case where these are not independent is tautological) and then using Theorem 2. We check

$$T_a R_\ell p = T_a p = p + a, \qquad R_m T_{r-p} p = R_m r = p + a,$$
$$T_a R_\ell q = T_a q' = q' + a, \qquad R_m T_{r-p} q = R_m(q' + a) = q' + a,$$
$$T_a R_\ell r = T_a r = r + a, \qquad R_m T_{r-p} r = R_m r' = r + a,$$

from the diagram.

There are several ways of deducing the following theorem from the results that have been proved already, the reader is urged to find some alternative proofs for himself.

**Theorem 6** Any isometry of $\mathbf{R}^2$ is the identity, a translation, a rotation, a reflection or a glide.

**Proof** Suppose first that f has a fixed point a. (A fixed point of f is a point a such that $fa = a$.) Corollary 1 to Theorem 4 says that, if a is taken to be 0, f is orthogonal. Hence f is represented by one of the matrices

$$\begin{bmatrix} \cos\alpha & -\sin\alpha \\ \sin\alpha & \cos\alpha \end{bmatrix} \quad \text{or} \quad \begin{bmatrix} \cos\alpha & \sin\alpha \\ \sin\alpha & -\cos\alpha \end{bmatrix}$$

according as f is direct or opposite. The first matrix represents a rotation with angle $\alpha$ and the second represents a reflection in the line $y = x \tan(\alpha/2)$.

Secondly, assume f has no fixed point and that $fa = b$. Let $\ell$ be the perpendicular bisector of ab, then $R_\ell fa = a$. By the above $R_\ell f$ is either a rotation about a or a reflection in a line through a. If $R_\ell f$ is reflection in the line m, $f = R_\ell R_m$ and $\ell$, m cannot meet, otherwise f has a fixed point, so $\ell$, m are parallel and f is a translation. The remaining case is when $R_\ell f = R(a,\alpha)$. Clearly $a \notin \ell$ otherwise a would be a fixed point of f. The rotation $R(a,\alpha)$ can be written as $R_m R_n$ and m can be chosen to be any line through a. Choose m to be parallel to $\ell$, then $f = R_\ell R_m R_n = T_a R_n$, where a is the vector perpendicular to $\ell$ and m of length twice the distance between $\ell$ and m. By the lemma $f = T_a R_n$ is a glide or a refection, and as it has no fixed point is must be a glide.

We can summarize all this in the following table which gives a classification of the non-identity elements of $\mathbf{I}(\mathbf{R}^2)$

| | Fixed Point? | |
|---|---|---|
| | Yes | No |
| Direct | Rotation | Translation |
| Opposite | Reflection | Glide |

**Isometries of $\mathbf{R}^3$**

The main examples are:

1. **Translations**
2. **Rotations** Let $\ell$ be a directed line in $\mathbf{R}^3$, $R(\ell,\alpha)$ denotes rotation about $\ell$ through angle $\alpha$. If $\ell$ is the z-axis, then $R(\ell,\alpha)$ has matrix

$$\begin{bmatrix} \cos\alpha & -\sin\alpha & 0 \\ \sin\alpha & \cos\alpha & 0 \\ 0 & 0 & 1 \end{bmatrix}$$

3. **Screws** A screw is the composition of a translation and a rotation. Let a be a vector parallel to the line $\ell$, then a typical screw is $T_aR(\ell,\alpha) = R(\ell,\alpha)T_a$.

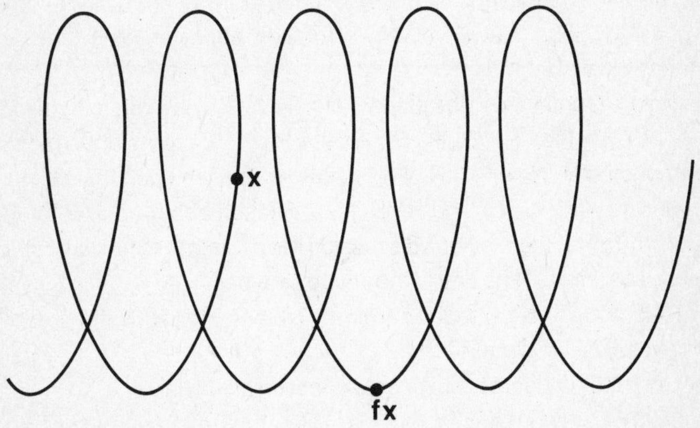


**Lemma** For any a, $\ell$, the composite $T_aR(\ell,\alpha)$ is a screw unless a is perpendicular to $\ell$.

**Proof** Write $a = a_1 + a_2$ where $a_1$ is parallel to $\ell$ and $a_2$ is perpendicular to $\ell$. Then $T_aR(\ell,\alpha) = T_{a_1}T_{a_2}R(\ell,\alpha)$. The transformation $T_{a_2}R(\ell,\alpha)$ is a rotation about an axis m that is parallel to $\ell$ (essentially, this is a two dimensional situation). Hence, $T_aR(\ell,\alpha)$ is the composition of a translation $T_{a_1}$ and a rotation about an axis parallel to $a_1$. Hence $T_aR(\ell,\alpha)$ is a screw unless $a_1 = 0$.

4. **Reflections** If H is a plane in $\mathbf{R}^3$, then $\mathbf{R}_H$ is reflection in H.
5. **Glides** If H is a plane in $\mathbf{R}^3$ and a is a vector parallel to H, then a typical glide is

$$T_a R_H = R_H T_a.$$

**Lemma** For any a, H, the composite $R_HT_a$ is a glide unless a is perpendicular to H, in which case it is a reflection.

**Proof** The cases when a is perpendicular to H and when a is parallel to H give a reflection and a glide respectively.

In general, let $a = a_1 + a_2$ where $a_1 \perp H$ and $a_2$ is parallel to H, then

$$R_H T_a = R_H T_{a_1} T_{a_2} = R_{H^1} T_{a_2}$$

where $H^1$ is a plane parallel to H. Hence this is a glide.

6. **Rotatory reflection** If H is a plane in $\mathbf{R}^3$ and $\ell$ is a line perpendicular to H, then a typical rotatory reflection is

$$R_H R(\ell,\alpha) = R(\ell,\alpha) R_H.$$

An important special case is when $\alpha = \pi$, this composite is called an **inversion** $I_a$, or a reflection in the point a. If $\ell \cap H = \{0\}$ then inversion is linear and its matrix is diagonal with three minus ones on the diagonal. Inversion at the point a is the composition of the reflections in any three mutually perpendicular planes meeting in a.

7. **A rotatory inversion** is the composition of a rotation about a line $\ell$ and inversion in a point a contained in $\ell$. Every rotatory inversion can be written as a rotatory reflection and vice versa.

Let $R(\ell,\alpha)I_a$ be a rotatory inversion. Write $R(\ell,\alpha)$ as $R_{H_1}R_{H_2}$ where $H_1$, $H_2$ are two planes meeting in $\ell$ at angle $\alpha/2$. Write $I_a$ as $R_{H_2}R_{H_3}R_{H_4}$ where $H_3$, $H_4$ are planes through a such that $H_2$, $H_3$ and $H_4$ are mutually perpendicular. Then $R(\ell,\alpha)I_a = R_{H_1}R_{H_3}R_{H_4}$. But $H_1$, $H_3$ meet in a line m that is perpendicular to $H_4$, hence $R(\ell,\alpha)I_a = R(m,\beta)R_{H_4}$, a rotatory reflection.

Now consider a rotatory reflection $R_H R(\ell,\alpha)$ where $\ell$ is perpendicular to H. Write this as $R_H R_{H_1} R_{H_2}$, then $H_1$ and $H_2$ are both perpendicular at the point $a = H \cap H_1 \cap H_2$. Then $R_H R(\ell,\alpha) = R_H R_{H_1} R_{H_3} R_{H_3} R_{H_2} = I_a . R_{H_3} R_{H_2}$. But $R_{H_3}R_{H_2}$ is a rotation about an axis that contains a, hence $R_H R(\ell,\alpha)$ is a rotatory inversion.

**Lemma** If $\ell$ is a line not contained in the plane H then $R_H R(\ell,\alpha)$ is a rotatory reflection.

**Proof** Let $f = R_H R(\ell,\alpha)$, then $f = R_H R_{H_1} R_{H_2}$ where H, $H_1$, $H_2$ are planes meeting in $a = \ell \cap H$ and H, $H_1$ are perpendicular. Now $I_a = R_{H_3} R_{H_1} R_H$ if $H_3$ is chosen such that H, $H_1$, $H_3$ are mutually perpendicular at a. So $f = I_a R_{H_3} R_{H_2}$ and $R_{H_3} R_{H_2}$ is a rotation about an axis through a. Hence f is a rotatory inversion and so by the previous discussion it is a rotatory reflection.

**Theorem 7** Every isometry of $\mathbf{R}^3$ is one of the above six types.

**Proof** We use Theorem 4 to show that an arbitrary isometry f of $\mathbf{R}^3$ is the product of at most four reflections and then analyse each case.

The cases where f is the product of zero or one reflection are trivial.

When f is the product of two reflections, $f = R_{H_1} R_{H_2}$, there are two cases: $H_1$, $H_2$ meet or $H_1$, $H_2$ are parallel. In the first case f is a rotation about the line where $H_1$, $H_2$ meet and in the second case f is a translation perpendicular to the planes through twice the distance between them.

When f is the product of three reflections, by the previous case f is $R_H T_a$ or $R_H R(\ell,\alpha)$. By the lemma on glides, $R_H T_a$ is always a glide. By the lemma on rotatory reflections, $R_H R(\ell,\alpha)$ is a rotatory reflection.

When f is the product of four reflections there are two cases. First, if f has a fixed point, then, by Theorem 4, f is a product of at most three reflections, hence f is a translation or a rotation. Secondly, if f has no fixed point one can find a so that $T_a f$ has a fixed point (choose any x and let $a = -f(x)$). Hence f is the composite of two translations, hence it is a translation or the composite of a translation and a rotation $T_a R(\ell,\alpha)$. Hence by the lemma on screws it is a screw.

**Corollary** A direct isometry of $\mathbf{R}^3$ that has a fixed point has a fixed line. This is often paraphrased as "every rotation has an axis".

We will now study the topology of $\mathbf{I}(\mathbf{R}^2)$ and the way that the four kinds of isometries define subsets of it. The description will be in terms of certain subsets of $\mathbf{R}^3$, and we digress to define them.

**Some subsets of R³**

We describe and name certain subsets of $\mathbf{R}^3$. Any two homeomorphic spaces will always be given the same name.

First a subspace of $\mathbf{R}^2$, the **circle**

$S^1 = \{\underline{x} \in \mathbf{R}^2 \mid \|\underline{x}\| = 1\}$.

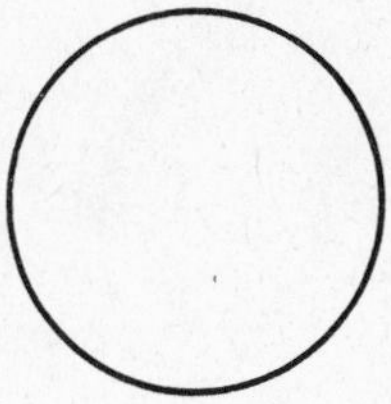

1. The **torus** $= \{\underline{x} \in \mathbf{R}^3 \mid (r-2)^2 + z^2 = 1 \text{ where } r^2 = x^2 + y^2\}$.

This is the surface obtained by revolving the circle $(x-2)^2 + z^2 = 1$ about the z-axis.

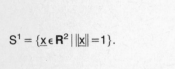

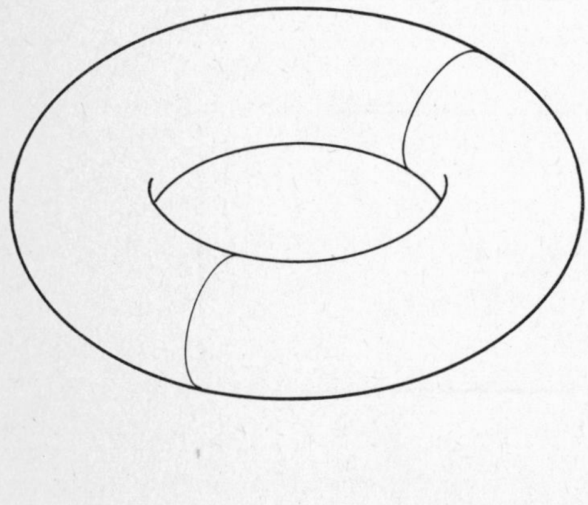

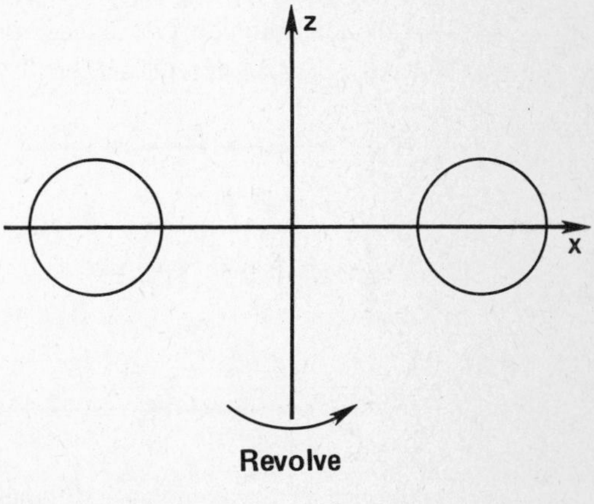


**Exercise** Show that $S^1 \times S^1$ is homeomorphic to the torus. The map is

$$(e^{i\theta}, e^{i\varphi}) \to ((2+\cos\varphi)\cos\theta, (2+\cos\varphi)\sin\theta, \sin\varphi).$$

The inverse map is

$$(x,y,z) \to ((x+iy)/|x+iy|, |x+iy|-2+iz)$$

Check that these maps are in fact inverses and note that they are both continuous.

2. **The solid torus** $= \{\underline{x} \in \mathbf{R}^3 \mid (r-2)^2 + z^2 \leqslant 1\}$

This is the torus together with its "inside", and is homeomorphic to $S^1 \times D^2$, where

$$D^2 = \{\underline{x} \in \mathbf{R}^2 \mid \|\underline{x}\| \leqslant 1\}.$$

The **open solid torus** is the "inside" of the solid torus, that is

$$\{\underline{x} \in \mathbf{R}^3 | (r-2)^2 + z^2 < 1\}$$

and is homeomorphic to $S^1 \times \mathbf{R}^2$ because $\mathbf{R}^2$ is homeomorphic to the open disc

$$\mathring{D}^2 = \{x \in \mathbf{R}^2 \mid \|x\| < 1\}.$$

3. The **Möbius band** M is the subset of the solid torus defined by
$x = r\cos\theta$, $y = r\sin\theta$, $z = (r-2)\tan\theta/2$.

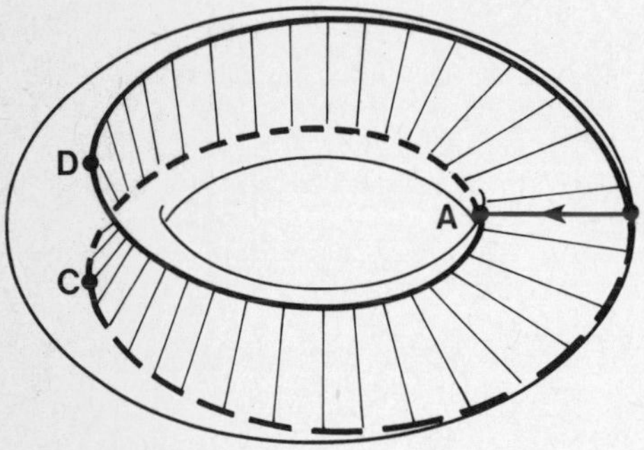


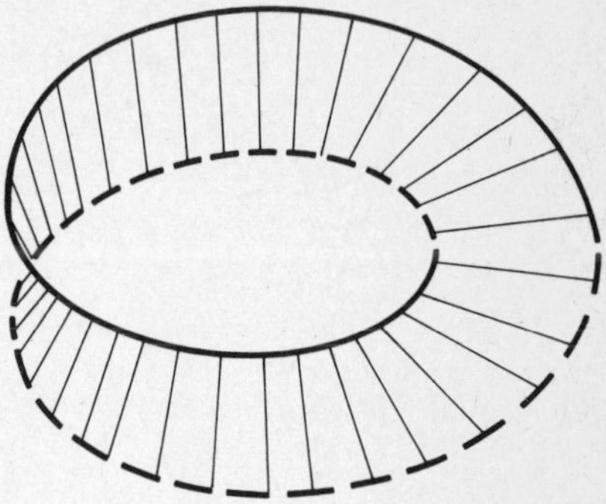

As we consider planes rotating around the z-axis – the angle $\theta$ increasing from 0 to $2\pi$, the Möbius band meets these planes in line segments of length 2 centred on the circle $x^2 + y^2 = 2$, $z = 0$ and the slope of these segments increases from 0 to $\pi$. If we cut this Möbius band along the line segment where $\theta = 0$ we get its familiar representation:

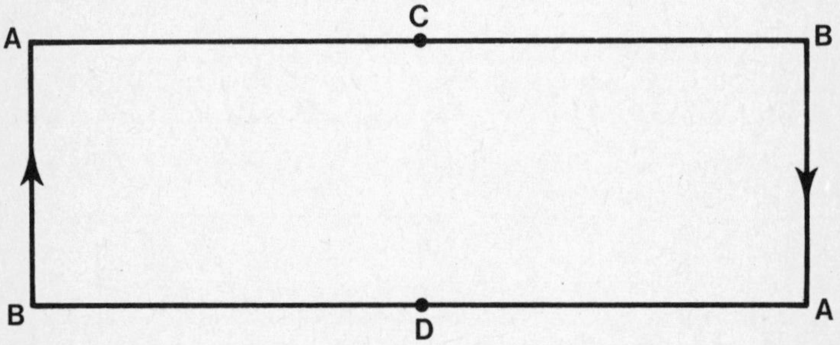


This notation means: take a rectangle and identify its opposite edges with a twist. More formally: Consider the set $[-1,1] \times [-1,1]$ and the equivalence relation

$$(x,y) \sim (u,v) \Leftrightarrow x = u \text{ and } y = v.$$
$$\text{or } x = 1, u = -1 \text{ and } y = -v.$$

Then the space we want is $[-1,1] \times [-1,1]/\sim$ with the metric $d([x],[y]) = \min\{d(x,y) | x \in [x], y \in [y]\}$.

The **open Möbius band** is $M \cap$ open solid torus.

**Exercise** Consider a Möbius band in $\mathbf{R}^3$ and "thicken it". Show that the resulting solid is homeomorphic to a solid torus.

Later we will need the following description of the open Möbius band.

**Lemma** The space $S^1 \times \mathbf{R}^1/(x,y) \sim (-x,-y)$ is homeomorphic to the open Möbius band.

**Proof** The metric on the quotient space $S^1 \times \mathbf{R}^1/(x,y) \sim (-x,-y)$ is given as above.
We cut $S^1 \times \mathbf{R}^1$ along $1 \times \mathbf{R}^1$ to give $[0,2\pi] \times \mathbf{R}^1$.
Note that the point $-x \in S^1$ becomes $x + \pi$ in $[0,2\pi]$.

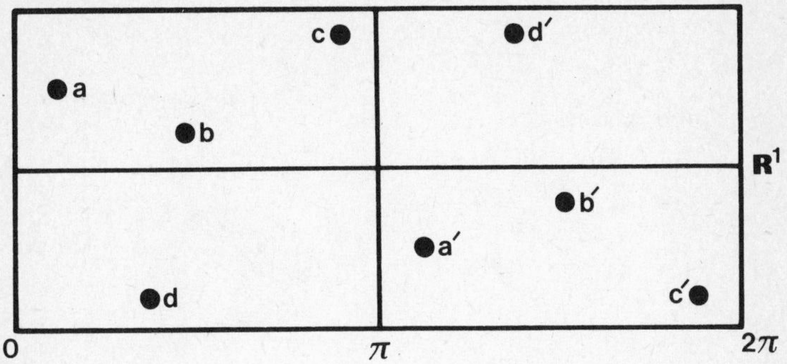


Under this equivalence relation, every point in the left hand half is identified with some point in the right hand half (a with a′ etc). Hence the quotient space is obtained from the left hand half by identifying points on its vertical edges appropriately.

Before we go on to study the topology of the space $\mathbf{I}(\mathbf{R}^2)$, we prove an interesting result about the space of all lines in $\mathbf{R}^2$. For ease of notation it is convenient to use complex numbers to represent points in $\mathbf{R}^2$. Every line in $\mathbf{R}^2$ is given by an equation of the form

$$Az + \bar{A}\bar{z} + B = 0 \text{ with } A \in \mathbf{C} \setminus \{0\} \text{ and } B \in \mathbf{R}.$$

The line $\ell x + my + n = 0$ corresponds to $2A = \ell - im$ and $B = n$.
The equation $\lambda Az + \lambda\bar{A}\bar{z} + \lambda B = 0$ determines the same geometric line (for $\lambda \in \mathbf{R} \setminus \{0\}$). By a suitable choice of $\lambda$ the equation can therefore be chosen to be $Az + \bar{A}\bar{z} + B = 0$ with $|A| = 1$ and $B \in \mathbf{R}$. The only remaining ambiguity is that $(A,B)$ and $(-A,-B)$ determine the same line. Hence the set of all lines in $\mathbf{R}^2$ can be identified with

$$S^1 \times \mathbf{R}/(A,B) \sim (-A,-B).$$

This gives the set of lines a metric and we have

**Theorem 8** The space of all lines in $\mathbf{R}^2$ is homeomorphic to the open Möbius band.

**Exercise** Show that the two lines represented by $(A_1,B_1)$ and $(A_2,B_2)$ are parallel if and only if $A_1 = \pm A_2$. Show that the distance between the parallel lines represented by $(A,B)$ and $(A,C)$ is $|B-C|$.

**Exercise** Show that the space of all directed lines in $\mathbf{R}^2$ is homeomorphic to $S^1 \times \mathbf{R}$.

**Corollary** The subspace of $\mathbf{I}(\mathbf{R}^2)$ consisting of reflections is homeomorphic to the open Möbius band.

In Theorem 5 is was proved that $\mathbf{I}(\mathbf{R}^2)$ is homeomorphic to $O(2) \times \mathbf{R}^2$. The group $O(2)$ is homeomorphic to the union of two disjoint circles. Hence $\mathbf{I}(\mathbf{R}^2)$ is homeomorphic to the union of two disjoint open tori, one consisting of the direct isometries and the other of the opposite isometries. These are the components of $\mathbf{I}(\mathbf{R}^2)$.

To examine this space in detail we again use the complex numbers. A translation is

$z \to z + b$ and a rotation about 0 is $z \to az$ with $|a| = 1$. The direct component consists of all transformations of the form $z \to az + b$ with $|a| = 1$ and so is homeomorphic to $S^1 \times \mathbf{R}^2$ (compare Theorem 5). The subset $1 \times \mathbf{R}^2$ consists of translations and its complement consists of rotations. The centre of $z \to az + b$ is at $b(1-a)^{-1}$ and it is a rotation through angle arg a. If the centre tends to infinity in such a way that b is constant we see that a translation is the limit of rotations. Of course this is clear in the diagram of $S^1 \times \mathbf{R}^2$.

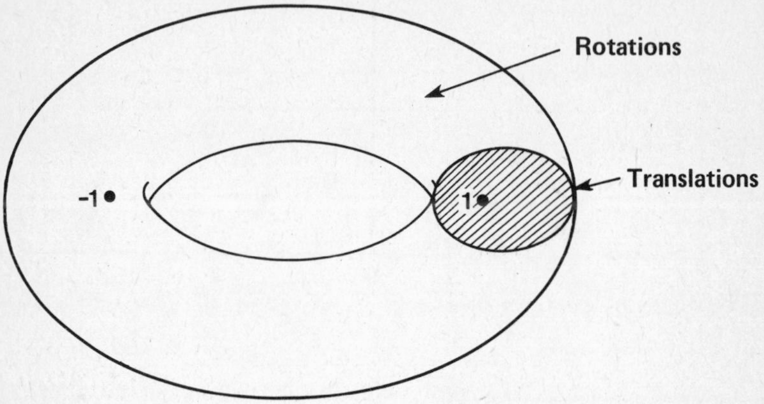


Now we consider the other component which is made up of all the opposite isometries. This component is a coset of the subgroup of direct isometries. The transformation $z \to \bar{z}$ is a reflection, and so this coset is made up of all the transformations of the form $z \to a\bar{z} + b$ with $|a| = 1$. Again $(a,b) \in S^1 \times \mathbf{R}^2$ gives an explicit homeomorphism between the component and $S^1 \times \mathbf{R}^2$. The reflections can be characterized as the opposite transformations of order 2 and so correspond to the set of (a,b) satisfying $a\bar{b} + b = 0$. By Theorem 8 this subset is a homeomorphic to the Möbius band, in fact it sits inside $S^1 \times \mathbf{R}^2$ in the way described on page 18 because $(e^{i\theta},b)$ satisfies $e^{i\theta}\bar{b} + b = 0$ if and only if $2\arg b = (\theta \pm \pi)$.

**Exercise** The transformation f: $z \to a\bar{z} + b$ leaves the set $F_{a,b} = \{z \in \mathbf{C} | z = a\bar{z} + b\}$ fixed. For most values of a, b with $|a| = 1$ the set $F_{a,b}$ is empty, in this case f is a glide. By subsituting $\bar{z} = \bar{a}z + \bar{b}$ into $z = a\bar{z} + b$ show that if $F_{a,b}$ is not empty then $a\bar{b} + b = 0$. Conversely, if $a\bar{b} + b = 0$ show that $F_{a,b}$ is non empty by checking that $\alpha + b/2 \in F_{a,b}$ where $\alpha^2 = a$. In the case $F_{a,b}$ is non-empty, $F_{a,b}$ is the line $Az + \bar{A}\bar{z} + B = 0$ where $A = \beta$, $B = -\beta b$ and $\beta^2 = -\bar{a}$.

The complement of this Möbius band is the set of glides. A glide is given by a reflection and a translation. The translation is through the vector $(a\bar{b}+b)/2$ and the reflection is

$$z \to a\bar{z} + (b - a\bar{b})/2.$$

As the length of the glide tends to zero the glide tends to a reflection, and this can be seen in the diagam of $S^1 \times \mathbf{R}^2$. The space of all glides is itself homeomorphic to an open torus. To see this cut $S^1 \times \mathbf{R}^2$ along $1 \times \mathbf{R}^2$ and remove the open Möbius band that is inside $S^1 \times \mathbf{R}^2$. We get the following object.

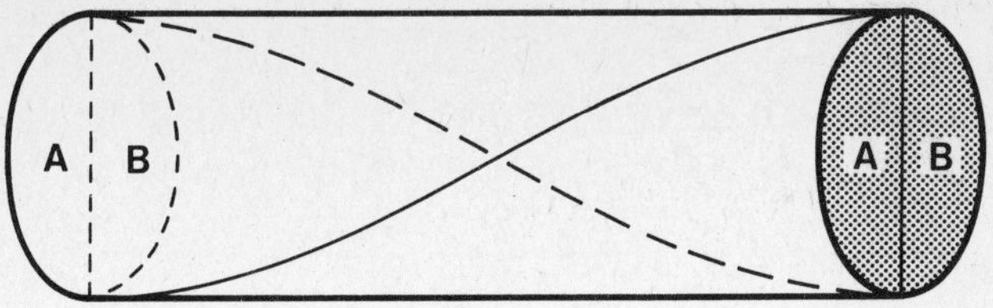


The 'twisted' line segment has been removed in each vertical section of the cylinder and the two half $\mathbf{R}^2$'s labelled A are to be identified one with the other as are the ones labelled B, as follows.

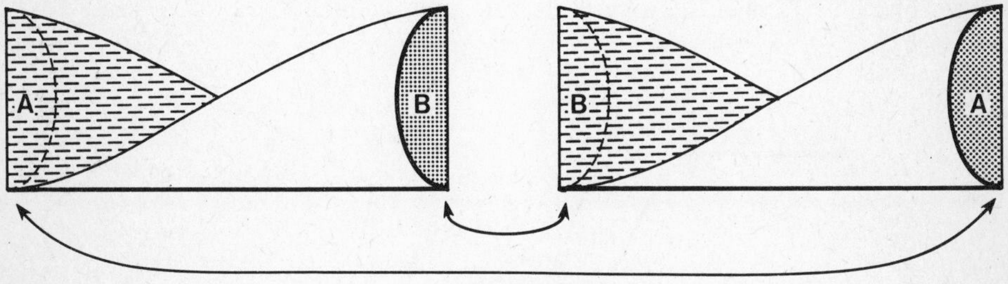


The resulting space is easily seen to be homeomorphic to the open solid torus.

**Exercise** Think of this $S^1 \times \mathbf{R}^2$ as being homeomorphic to a Möbius band made out of thick cardboard. The process described above is then equivalent to cutting the Möbius band down the middle. This leaves a solid that is homeomorphic to a thick annulus and this is an open solid torus.

**Finite Groups of Isometries**

If $X \subset \mathbf{R}^n$ contains $(n+1)$ independent points its **symmetry group** $S(X)$ is the subgroup of $\mathbf{I}(\mathbf{R}^n)$ consisting of those isometries f such that $fX = X$. It is also the group of isometries of X (remember the exercise after Theorem 3). One may need to consider subsets X of $\mathbf{R}^n$ that do not contain $(n+1)$ independent points, there are two alternatives: the first is to consider isometries of $\mathbf{R}^n$ that send X to itself but there may be many of these that are the identity on X, the second alternative is to regard X as a subset of the affine space A spanned by it and considering the isometries of A that send X to itself.

**Example 1** If $\ell$ is a line in $\mathbf{R}^2$, it does not contain 3 independent points and the subgroup of $\mathbf{I}(\mathbf{R}^2)$ that preserves $\ell$ contains a normal subgroup with two elements whose quotient is $\mathbf{I}(\mathbf{R}^1)$.

**Example 2** If $X \subset \mathbf{R}^2$ is $\{(x,0)|x \in \mathbf{R}\} \cup \{(0,y)\,|-1 \leq y \leq 1\}$

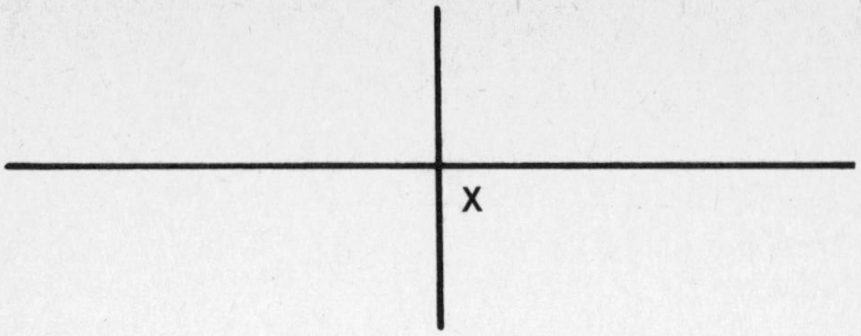


then $S(X) = \mathbf{Z}/2 \times \mathbf{Z}/2$, the three non-trivial elements being $R_x$, $R_y$, $R(0,\pi)$. (Here $R_x$ denotes reflection in the x-axis.)

**Example 3** If $P_n$ is a regular n-gon with centre 0 (say). $S(P_n)$ is the **dihedral group** $D_n$ having 2n elements. Any $f \in S(P_n)$ keeps 0 fixed, hence $S(P_n) \subset O(2)$. It contains the elements $S = R(0,2\pi/n)$ and $T = R_\ell$ where $\ell$ is a line containing 0 and a vertex of $P_n$.

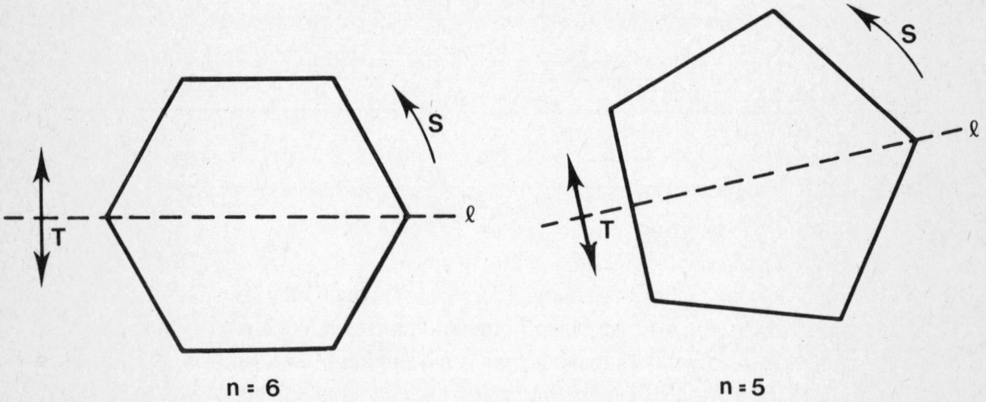


The element S generates a normal cyclic subgroup $\{S\}$ of order n and T has order 2. The transformation S is direct and T is opposite, in fact $\{S\} = S(P_n) \cap SO(2)$ and so has index 2 in $S(P_n)$. So $S(P_n)$ is made up of $\{S\}$ and its coset $T\{S\}$. All the elements of $T\{S\}$ are reflections in lines through 0 and either through a vertex or through the mid point of an edge of $P_n$. The group $S(P_n)$ is generated by S and T with the relations $S^n = 1$, $T^2 = 1$, $TST = S^{-1}$.

The dihedral groups in effect describe all the finite subgroups of $\mathbf{I}(\mathbf{R}^2)$:

**Theorem 9** Every finite subgroup of $\mathbf{I}(\mathbf{R}^2)$ is either cyclic or dihedral.

**Proof** Let $G \subset \mathbf{I}(\mathbf{R}^2)$ be a subgroup of order n. Choose any point $a \in \mathbf{R}^2$. Its orbit under the action of G, $\text{Orb}(a) = \{ga|g \in G\}$, is a finite subset of $\mathbf{R}^2$. Let $c = (\sum_{g \in G} ga)/n$ be the centroid of the set Orb(a). As the centroid c of a finite set $X \subset \mathbf{R}^2$ is a particular affine combination of the points of X and an isometry is an affine map, the point fc is the same affine combination of the points of fX, hence fc is the centroid of fX. If $g \in G$, then $g\text{Orb}(a) = \text{Orb}(a)$, hence $gc = c$ for each $g \in G$. This proves that G is a subgroup of $O(2)_c$ the group of orthogonal transformations centred at c. First consider the direct subgroup of G, $G_d = G \cap SO(2)$. This subgroup consists of a finite number of rotations. If S is the rotation with least angle amongst those of $G_d$, it is easy to see that

$G_d$ is the cyclic group generated by S. If $G = G_d$ one is finished; otherwise $G_d$ has index 2 in G, because SO(2) has index 2 in O(2). The set $O(2) \setminus SO(2)$ consists entirely of reflections, suppose $T = R_\ell$ is a reflection in G; then G is generated by S and T, and it is easy to check that $TST = S^{-1}$ (this holds for any rotation S and reflection T). Hence the group G is dihedral.

This theorem shows that there is an intimate connection between the finite isometry groups of the plane and regular figures in the plane. We will now proceed to prove the corresponding result for $\mathbf{R}^3$; but first here is an alternative to the centroid argument in the above proof.

**Exercise** Let G be a subgroup of $\mathbf{I}(\mathbf{R}^2)$. Show that G is infinite if it contains both a non-trivial rotation about a point a and a reflection in a line $\ell$ such that $a \notin \ell$ (you should use the Lemma on page 13). Hence show that G is infinite if it contains reflections in three non-concurrent lines. Deduce that if G is finite then there is a point fixed under all the elements of G.

## The Platonic Solids

We start with some definitions. A subset X of $\mathbf{R}^n$ is **convex** if for every pair of points $x, y \in X$ the line segment joining them lies entirely within X, that is, for each t with $0 \leqslant t \leqslant 1$, the point $tx + (1-t)y$ is in X. That is, X is closed under taking **non-negative** affine combinations. A subset of $\mathbf{R}^n$ defined by a finite set of linear inequalities is clearly convex and is called a **convex polyhedron**. (A finite union of convex polyhedra is called a **polyhedron**.) So a convex polyhedron X in $\mathbf{R}^3$ is bounded by planes, a two-dimensional subset of X that is in such a plane is called a **face** of X, the intersection of two faces an **edge** and the intersection of two edges a **vertex**. (We will always assume that a convex polyhedron X has nonempty interior thus ensuring that X is really 'solid' and does not lie in any affine hyperplane of $\mathbf{R}^3$.) A face is homeomorphic to a closed disc, an edge to a closed interval and a vertex is a point.

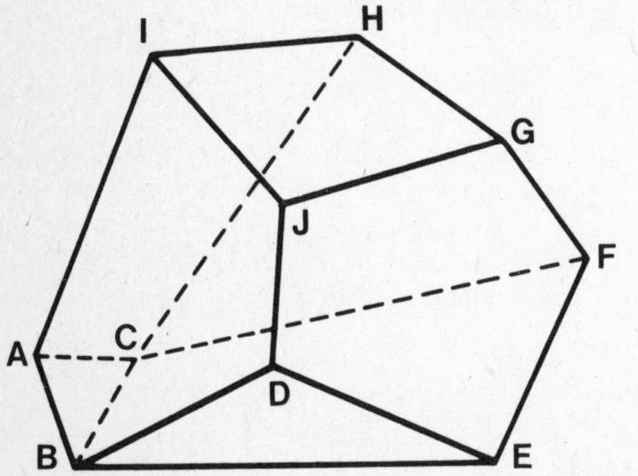


Faces: ABC, ACHI, ABDJI, BCFE, BDE, CFGH, DEFGJ, GHIJ
Edges: AB, AC, AI, BC, BD, BE,CF, CH, DE, DJ, EF, FG, GH, GJ, HI, IJ

A polyhedron is called **regular** if all its faces, edges and vertices are identical to each other. By saying that two vertices are identical one means that there are the same number of edges at each vertex and that the angles between them are the same. A convex regular polyhedron is also called a **Platonic solid** because they were studied by the Platonic school and played a role in their philosophy.

**Exercise** The definition of a Platonic solid given here uses "overkill". Find how many of the assumptions can be dropped because they are consequences of others.

A Platonic solid is described by the number of its faces; 'hedron' means 'seat' in Greek and tetra etc. are the Greek words for four, . . . :

**Tetrahedron**

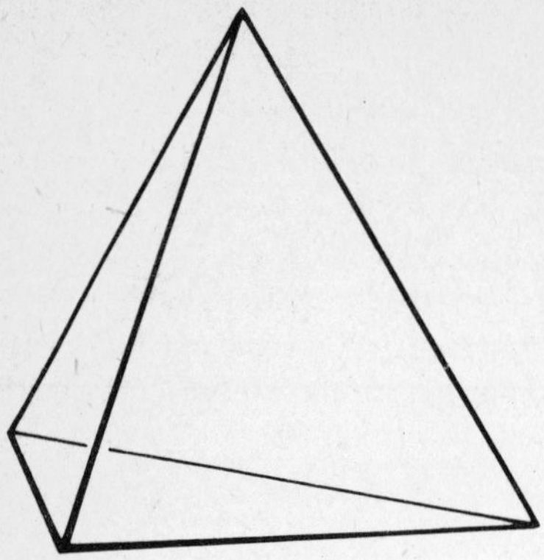

4 vertices each with 3 edges at $\pi/3$
6 edges
4 faces each an equilateral triangle

**Cube (or Hexahedron)**

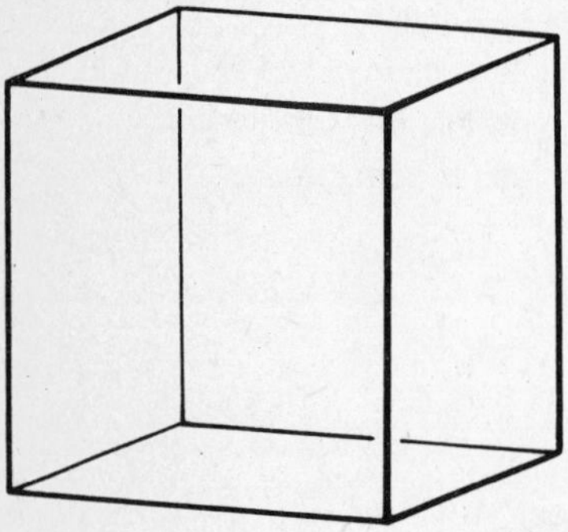

8 vertices each with 3 edges at $\pi/2$
12 edges
6 faces each a square

**Octahedron**

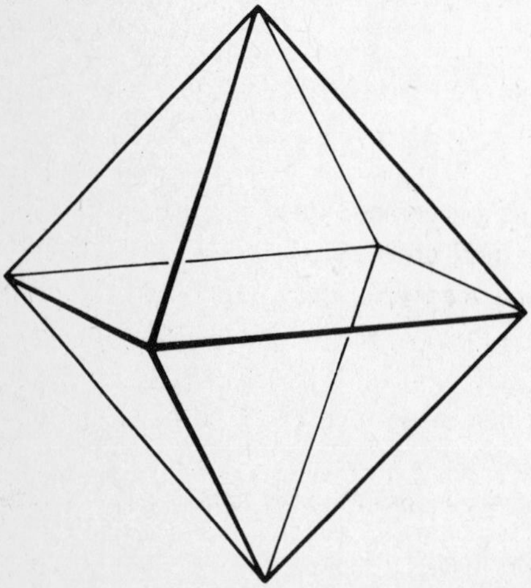

6 vertices each with 4 edges at $\pi/3$
12 edges
8 faces each an equilateral triangle

**Dodecahedron**

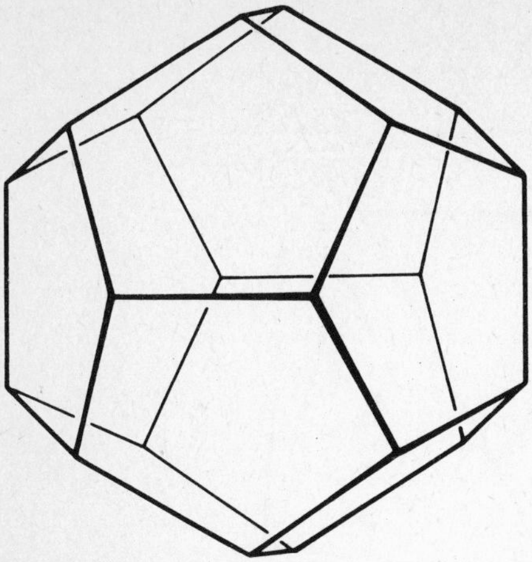

20 vertices each with 3 edges at $3\pi/5$
30 edges
12 faces each a regular pentagon

**Icosahedron**

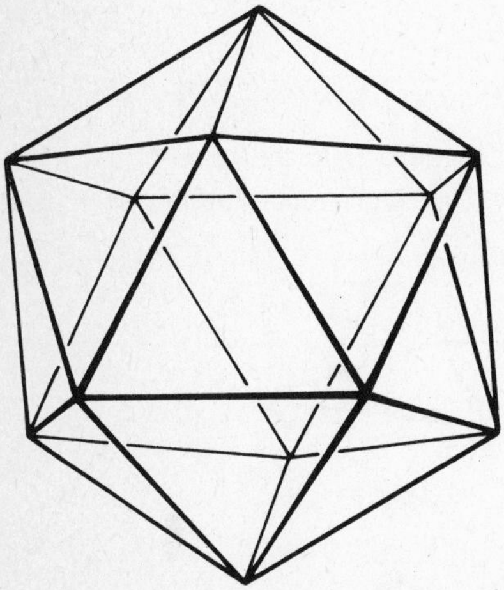

12 vertices each with 5 edges at $\pi/3$
30 edges
20 faces each an equilateral triangle

The first three are simple enough to understand and models of them are easily constructed. I will explain briefly how to construct the other two.

The icosahedron is obtained by first taking a pyramid on a regular pentagon A.

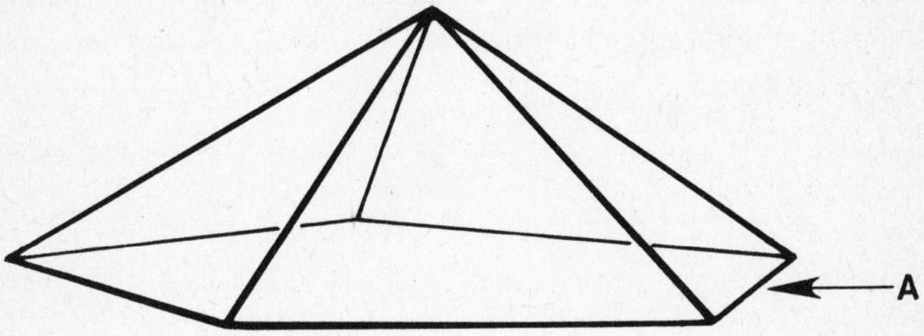

Take five more equilateral triangles B one on each edge of the pentagon A, interspace these with five more such triangles C, forming a drum.

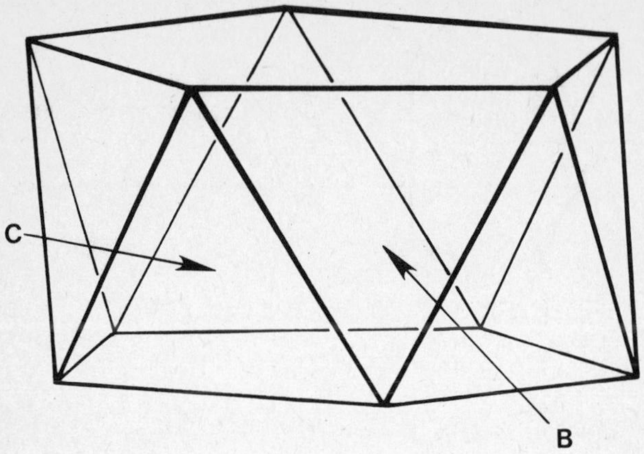


Finally complete the icosahedron by adding another pyramid D to the base of the drum. By construction, all the faces will be equilateral triangles and one can deduce that the resulting object is Platonic.

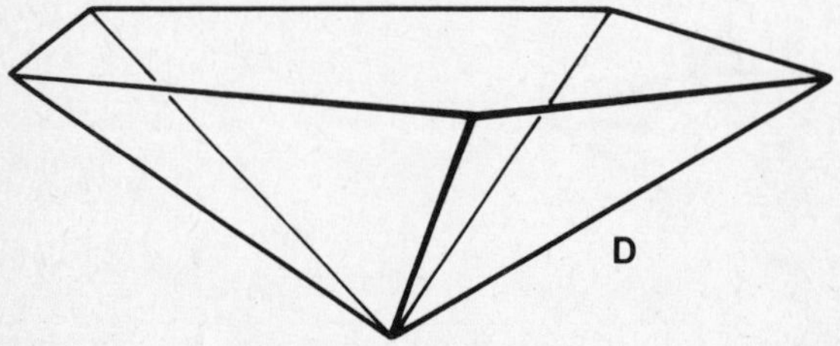


The dodecahedron is obtained by taking a regular pentagon A and adding to it 5 regular pentagons B, as shown, to form a upside down bowl. If one takes another such bowl, the two fit together to form a dodecahedron.

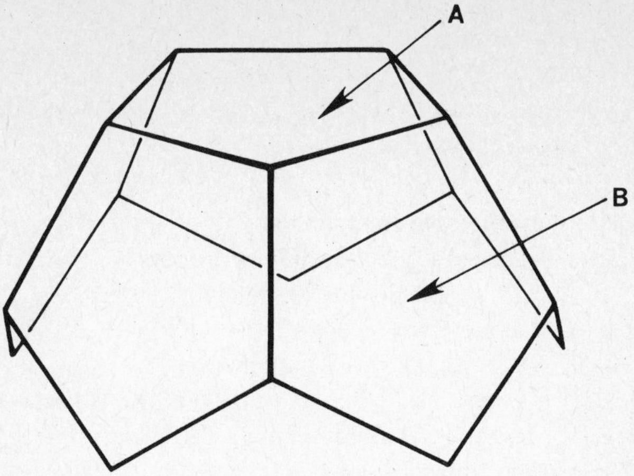


**Exercise** Construct models of these solids using cardboard and sticky tape (or rods or . . . ).

**Duality**

If one takes a Platonic solid, joins the mid points of adjacent faces by a new edge and fills in the resulting solid, one obtains the **dual** solid which is again Platonic. If this process is done twice it is easy to see that one recovers the original Platonic solid (or at least a smaller version of it). The tetrahedron is self dual, the cube and octahedron are dual, as are the dodecahedron and icosahedron. If one tabulates the number of vertices, edges and faces one sees the duality in the table:

| | Vertices | Edges | Faces |
|---|---|---|---|
| Tetrahedron | 4 | 6 | 4 |
| Cube | 8 | 12 | 6 |
| Octahedron | 6 | 12 | 8 |
| Dodecahedron | 20 | 30 | 12 |
| Icosahedron | 12 | 30 | 20 |

Note that Euler's formula $V-E+F=2$ holds in all five cases. This formula holds for any polyhedron that is homeomorphic to the 3-dimensional disc $D^3 = \{x \in \mathbf{R}^3 \mid \|x\| \leqslant 1\}$ so in particular it is true for convex polyhedra. (Algebraic topology gives generalisations of this formula to polyhedra in $\mathbf{R}^n$.)

As was known to the ancient Greeks these five are the only Platonic solids:

**Theorem 10** There are precisely five Platonic solids.

**Proof** We have already constructed the five so we need only prove that there are no more.

Suppose that r faces meet at each vertex and that each face is a regular n-gon. It is clear that each of r and n is at least 3. The sum of the angles at a vertex is less than $2\pi$, and each angle is $(n-2)\pi/n$, being the angle of a regular n-gon. Hence we see that

$$(r(n-2)\pi)/n < 2\pi$$

from which one deduces that

$$(r-2)(n-2) < 4.$$

It is easy to see that any integral solution of this inequality with r, $n \geqslant 3$ must have r or $n = 3$, so the only solutions are

$$(r,n) = (3,3), (3,4), (3,5), (4,3) \text{ and } (5,3).$$

If one knows the shape of a face (that is n) and how many faces meet at a vertex (that is r) then there is only one possible solid with that particular (r,n).

**The Symmetry Groups of the Platonic Solids.**

If X is a Platonic solid whose centre is at 0 its **symmetry group** is

$$S(X) = \{f \in O(3) \mid fX = X\}$$

and its **rotation group** (or direct symmetry group) is

$$S_d(X) = \{f \in SO(3) \mid fX = X\}.$$

The rotation group $S_d(X)$ is a normal subgroup of index 2 in S(X). (Of course, for some subsets $X \subset \mathbf{R}^3$ one may have $S(X) = S_d(X)$ but this is not the case if X is Platonic.)

We will use the following (standard) notation:

$S_n$ is the **symmetric group on n letters** i.e. Bij($\{1,2,\ldots,n\}$). This group has n! elements called **permutations**. An element of $S_n$ that merely interchanges two letters (say i and j) and leaves the others fixed is called a **transposition** and is denoted by (ij) in the cycle notation. It is standard and easy to show that every element of $S_n$ can be written as a product of transpositions.

**Exercise** Show that any $\sigma \in S_n$ can be written as a product involving only the (n–1) transpositions

$$(1,2), (2,3), \ldots, ((n-1)n).$$

The group $S_n$ has an important (normal) subgroup of index 2 called the **alternating group** $A_n$. It consists of those permutations that can be written as a product of an **even** number of transpositions. It is not quite obvious that this is a well defined condition on a permutation (and hence that the set of all such permutations forms a subgroup of $S_n$). Perhaps the easiest way to see this is to introduce variables $x_1, x_2, \ldots, x_n$ and make $\sigma \in S_n$ act on the polynomial ring $\mathbf{Z}[x_1, x_2, \ldots, x_n]$ by $\sigma(x_i) = x_{\sigma(i)}$ then

$$A_n = \{\sigma \in S_n \mid \sigma f = f \text{ where } f = \prod_{i<j} (x_i - x_j)\}.$$

As an alternative one has:

**Exercise** Define sign: $S_n \to \{\pm 1\}$ by

$$\text{sign}(\sigma) = \prod_{i<j} (\sigma(i) - \sigma(j))/(i-j)$$

Verify that i) sign (transposition) $= -1$

ii) sign is a homomorphism.

Deduce that $A_n = \text{Ker(sign)}$.

**Note** We will now assume that the reader is quite familiar with the above facts about permutations, he should therefore make sure that he understands and can prove them.

Let T denote the **regular tetrahedron**, we have $S(T) \cong S_4$ and $S_d(T) \cong A_4$:

By labelling the vertices of T with the numbers 1, 2, 3, 4 and because any $f \in S(T)$ must send a vertex to a vertex one obtains a homomorphism $\alpha: S(T) \to S_4$.

First, we show that $\alpha$ is onto. Because $S_4$ is generated by transpositions, it is enough to show that any transposition is in the image of $\alpha$. We show that (12) is in the image of $\alpha$, the proof for the others is identical. Consider the plane H through the vertices 3, 4 and through the mid point $m$ of the edge 12, H is the perpendicular bisector of 12, hence the reflection $R_H$ is such that $\alpha R_H = (12)$.

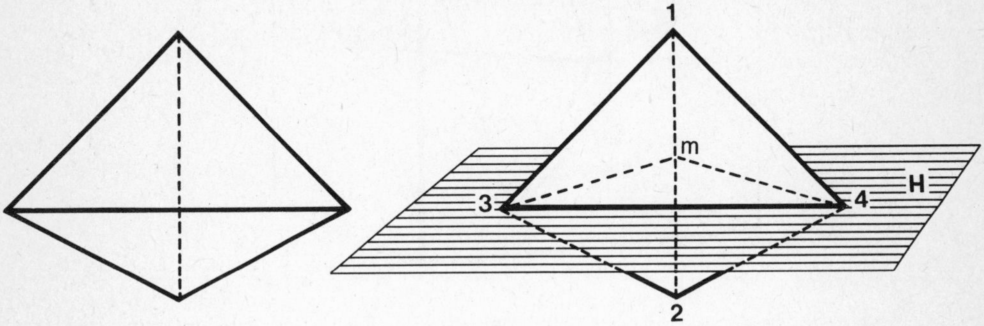


The homomorphism $\alpha$ is also injective because, by Theorem 2, any isometry of $\mathbf{R}^3$ is uniquely determined by the images of four affinely independent points (i.e. four non coplanar points). Hence $\alpha$ is an isomorphism.

It is now apparent, because a reflection in a plane is mapped by $\alpha$ to a transposition, that every direct isometry of T is mapped to an even permutation of $\{1,2,3,4\}$, that is, the image of $S_d(X)$ under $\alpha$ is contained in $A_4$. As $S_d(X)$ has index 2 in S(X) and $A_4$ has index 2 in $S_4$, $\alpha$ induces an isomorphism: $S_d(X) \to A_4$.

The other Platonic solids have a common feature – they are all **centrally symmetric**, that is the map J: $\mathbf{R}^3 \to \mathbf{R}^3$ given by $Jx = -x$ is a symmetry. This map J, often called **central inversion**, commutes with every linear transformation of $\mathbf{R}^3$. The group O(3) is isomorphic (as a group) to $SO(3) \times \{\pm 1\}$, the map being

$$A \to (A,1) \text{ if } \det A = 1$$
$$A \to (JA,-1) \text{ if } \det A = -1 \qquad (\det JA = 1 \text{ in this case}).$$

As J commutes with any A, this map is a homomorphism and its inverse is given by

$$(A,1) \to A, \qquad (A,-1) \to JA.$$

If the solid X is centrally symmetric, it is now easy to see that $S(X) \cong S_d(X) \times \{\pm 1\}$.

Another feature comes from duality. If X and Y are dual, then any symmetry of X is a symmetry of Y and vice versa, so that we have an isomorphism $S(X) \cong S(Y)$.

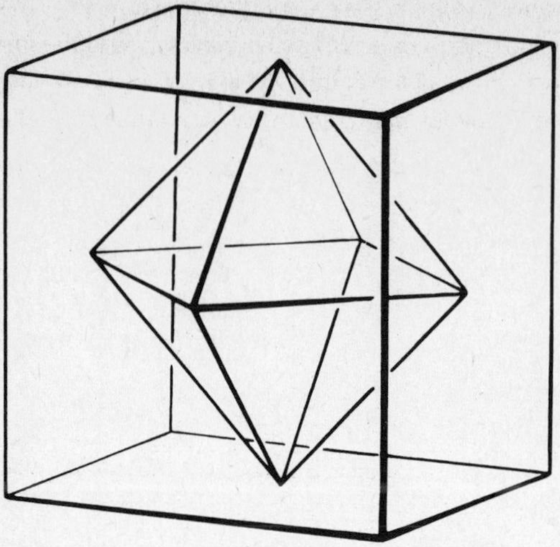

It is clear from these remarks on central symmetry and duality that if one determines the rotation groups $S_d(X)$ of the cube and the dodecahedron then all other symmetry groups are easy to find from these.

Let C denote the **cube**, then its rotation group $S_d(C)$ is isomorphic to $S_4$. To show this one must find four geometric objects in C that are permuted by the rotations of C. The four objects that we will use are the four diagonals of the cube, labelling them 1, 2, 3, 4.

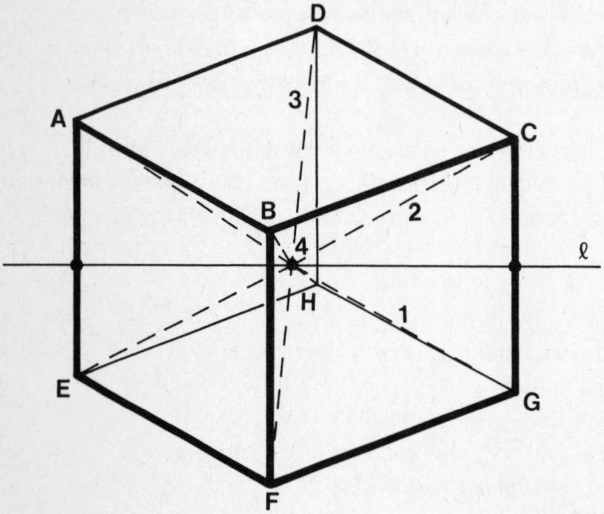


Diagonal 1: AG
2: CE
3: DF
4: BH

Consider the map $\alpha: S_d(C) \to S_4$ obtained in this way. To show $\alpha$ is onto it is enough to find a rotation f of C such that $\alpha(f) = (12)$, the other cases are identical. Consider the plane containing the diagonals 1 and 2. It contains two edges of C, namely AE and CG; let $\ell$ be the line joining the mid points of these edges. If f denotes the rotation through angle $\pi$ about $\ell$, one has that $fA = E$, $fC = G$, $fB = H$ and $fD = F$, and so $f^2 =$ identity. It is now clear that $\alpha f = (12)$ as required.

To prove that $\alpha$ is injective we will use the "pigeon-hole principle" (any map between two finite sets with the same number of elements is a bijection if it is onto). The group $S_4$ has $4! = 24$ elements and we will now check that $S_d(C)$ also has 24 elements. For any pair of vertices P, Q of the cube C, there is a rotation of C that sends P to Q (one can write down such a rotation for each case) and there are precisely three rotations that fix a given vertex. Hence there are 24 elements in $S_d(C)$. In the language of groups acting on sets, the group $S_d(C)$ acts on the cube, the set of 8 vertices is an orbit and the stabilizer of a vertex has 3 elements.

**Exercise** Prove that the points on the surface of a cube have orbits with either 24, 12, 8 or 6 elements under the action of $S_d(C)$. Describe geometrically the points having these different orbit sizes.

Let D denote the **docahedron**. We show that its rotation group $S_d(D)$ is isomorphic to $A_5$, a group which has 60 elements. To do this one must find five geometric objects associated with the dodecahedron such that they are permuted by the elements of $S_d(D)$. Five such objects are the cubes that are inscribed in the dodecahedron. The edges of these cubes are diagonals of the faces of the dodecahedron. Here are two descriptions of these cubes.

**First description**

Remember that a diagonal of a regular pentagon, with edge length 1, has length

$$\tau = (1+\sqrt{5})/2 = 2\cos(\pi/5).$$

The number $\tau$ is called the **golden ratio**, and is the positive root of $x^2 - x - 1 = 0$. One way of proving this is to consider the diagram.

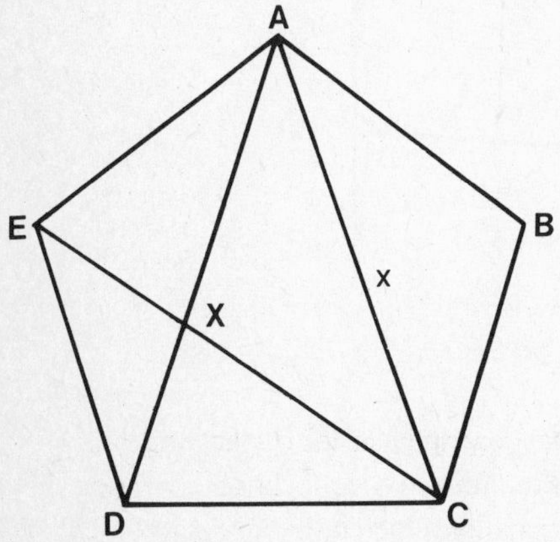


Clearly $AX = CX = AB = 1$ and $AC = EC = x$ so $EX = x - 1$. By similar triangles $AX/AC = EX/ED$ which gives the result:

$$1/x = (x-1)/1 \text{ so } x^2 = 1 + x.$$

Now consider a cube whose edge length is $\tau$. On a face of the cube we construct a 'tent' as follows: two faces are isosceles triangles with sides $\tau$, 1, 1 and the other faces are trapezoids with

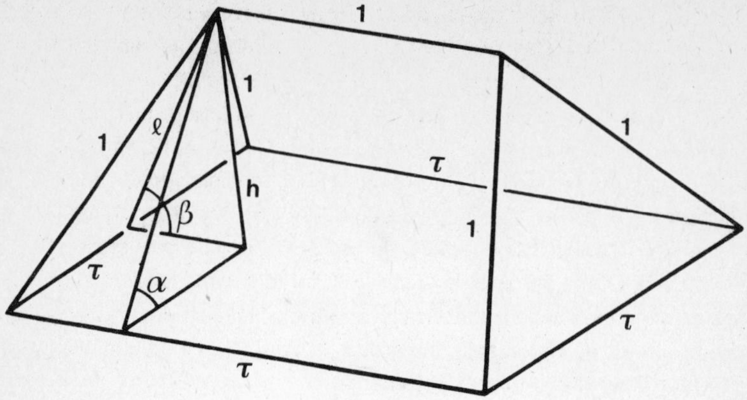


sides $\tau$, 1, 1, 1 as shown. The trapezoid and triangle make angles $\alpha$, $\beta$ respectively with the base. We will show that $\alpha + \beta = \pi/2$.

Let $\ell$ be the height of the triangle and h the height of the tent. These are given by the equations

$$1 = \ell^2 + (\tau/2)^2, \qquad \ell^2 = h^2 + ((\tau-1)/2)^2.$$

Substituting gives us $1 = h^2 + \tau^2/2 - \tau/2 + 1/4$. But $\tau^2 - \tau = 1$, so $h = 1/2$. We now deduce that $\tan\alpha = 2h/\tau = 1/\tau$ and $\tan\beta = 2h/(\tau-1) = 1/(\tau-1)$. So $\tan\alpha \tan\beta = 1$, and therefore $\alpha + \beta = \pi/2$. This shows that when we put two tents on adjacent faces of the cube as follows:

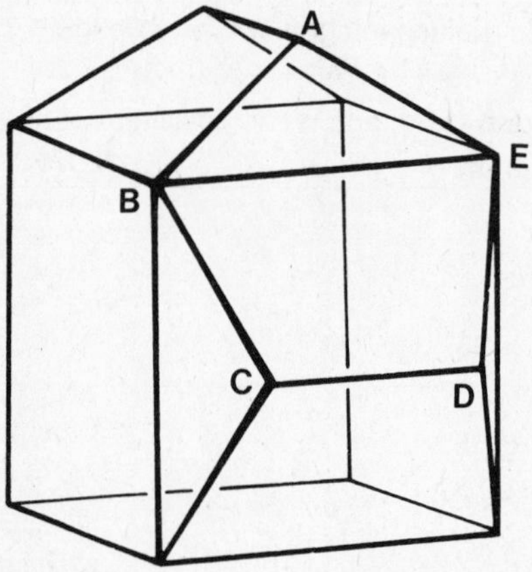


the pentagon ABCDE is planar.

**Exercise** Complete this argument to give another description of the dodecahedron.

In this way we have inscribed a cube of edge length $\tau$ in a dodecahedron of edge length 1 in such a way that each edge of the cube is a diagonal of a face of the dodecahedron. By starting with each of the five diagonals of a face of the dodecahedron one can obtain five different cubes inscribed in the dodecahedron. These are obtained from each other by applying 1, R, $R^2$, $R^3$, $R^4$, where R denotes rotation through angle $2\pi/5$ about an axis perpendicular to a face of the dodecahedron. Each cube contains two tetrahedra:

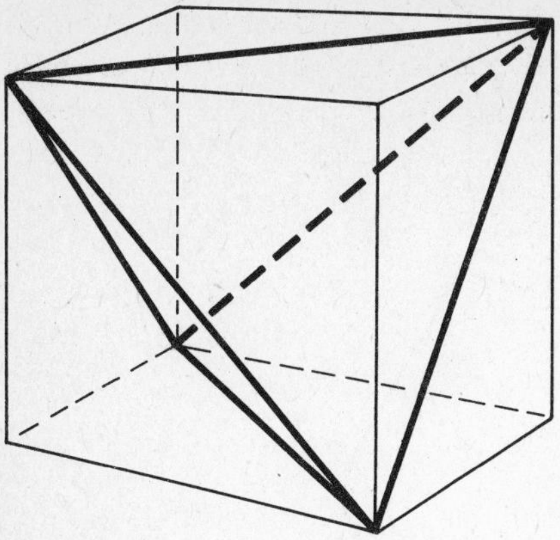

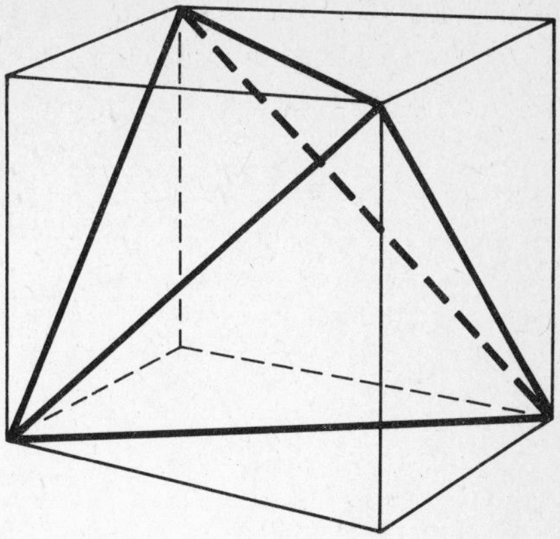

If one considers one of these and its images under the group $\{R\}$ one easily sees that the images of the four vertices are all different and so are all the 20 vertices of the dodecahedron. This leads to the

**Second description**

Consider the dodecahedron with one of its faces horizontal. Its vertices are then divided into four sets of five, each set at a different level. First label the vertices of the top face cyclically by 1, 2, 3, 4, 5. Secondly, label the vertices of the next level. Consider a particular vertex P at this second level, P is joined by an edge to one vertex, say, 5 of the top face and by a chain of two edges to two other vertices 1, 4 of the top face;

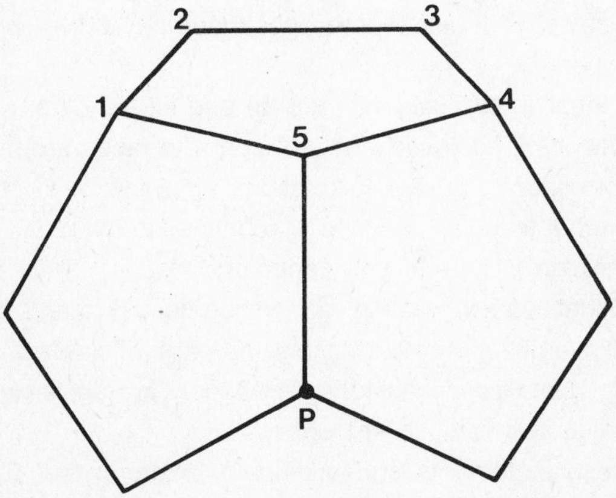


P is given one of the other labels, that is, 2 or 3 (there is a choice here but nowhere else, the two choices essentially differ only by a reflection). The other vertices of this level are then labelled cyclically in the same direction as the labelling of the top face. So far, the top face has been labelled completely and each of the five faces touching it has four of its vertices labelled. It is easy to check that for each of these five faces the four labels already assigned are different. The vertices at the next level are now labelled so that their five faces have vertices all with different labels. Finally one must label the vertices of the bottom face; each face touching it has three vertices already labelled:

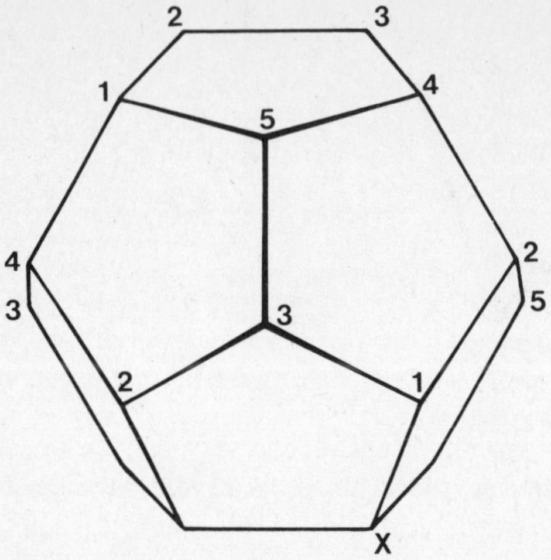


The vertex marked X cannot be labelled with 1, 2, 3 or 5 otherwise a face would have two vertices with the same label, so it must be labelled 4. This procedure works to give a complete labelling. Inspection shows that the four vertices with the same label form the vertices of a tetrahedron. Thus one has a system of five tetrahedra inscribed in D. There is another system of five tetrahedra inscribed in D obtained from the constructed system by a reflection. Alternatively, they would have been constructed if one had made the other choice at the second stage of the labelling procedure.

**Exercise** Colour the faces of your model of the icosahedron with five colours corresponding to the (dual of the) above procedure.

The twelve faces are labelled cyclically by (12345) and all the other 5-cycles obtained from (12345) by even permutations. The cyclic labels of opposite faces are mutual inverses. These twelve 5-cycles in $S_5$ correspond under $\alpha$ to rotations through $\pm 2\pi/5$ about the centre of the corresponding face. The other twelve 5-cycles in $S_5$ (obtained from (12345) by odd permutations) correspond to rotations through $\pm 4\pi/5$ about the centre of the faces.

**Note** It is a familiar fact that two permutations are conjugate in $S_n$ if and only if their expressions as the product of disjoint cycles involve cycles of the same length. This is not quite true for $A_n$: cycles of length n (for n odd) are in two distinct conjugacy classes. This fact is clear in $S_d(D)$ – rotations through $\pm 2\pi/5$ are not conjugate to rotations through $\pm 4\pi/5$.

**Exercise** Check all these statements and use them to prove that O(3) has no subgroup isomorphic to $S_5$. Find a subgroup of O(4) that is isomorphic to $S_5$.

To continue with the proof, one can check that the various rotations through $\pm 2\pi/3$ about the vertices give each of the (twenty) 3-cycles in $A_5$, and that the rotations through $\pi$ about the mid points of the edges give the (fifteen) elements of order 2 in $A_5$ (such as (12)(34)).

We have proved that $\alpha: S_d(D) \to A_5$ is onto. It is easy to check that $S_d(D)$ has 60 elements and so $\alpha$ is an isomorphism.

**Exercise** Show directly that there are elements of orders 3 and 5 in the image of $\alpha$ and that there is a subgroup of order 4 in the image of $\alpha$. Deduce that $\alpha: S_d(D) \to A_5$ is onto.

Summarizing we have the following.

**Table of Symmetry Groups**

| | Direct | Full |
|---|---|---|
| Tetrahedron | $A_4$ | $S_4$ |
| Cube, Octahedron | $S_4$ | $S_4 \times \{\pm 1\}$ |
| Dodecahedron, Icosahedron | $A_5$ | $A_5 \times \{\pm 1\}$ |

**Exercise** Prove directly that the number of elements in the rotation group of a Platonic solid is twice the number of its edges.

## Finite Groups of Rotations of $\mathbf{R}^3$

Now we will prove the analogue of Theorem 9 in the case of $\mathbf{R}^3$. We need the following result. (Remember that SO(n) is the group of orthogonal $n \times n$ matrices with determinant 1.)

**Lemma** If $A \in SO(n)$ and n is odd, then 1 is an eigenvalue of A.

**Proof** We show that det(A–I) is singular.

$$\det(A-I) = \det(A-I).\det A^t = \det(AA^t - A^t)$$
$$= \det(I - A^t) = \det(I-A)^t = \det(I-A) = (-1)^n \det(A-I).$$

But n is odd so $\det(A-I) = 0$.

**Note** This lemma is not true for all $A \in O(n)$, for example $A = -I$.

**Corollary 1** "Every rotation has an axis" – that is, if $A \in SO(3)$ then there is an orthogonal matrix P such that

$$P^{-1}AP = \begin{bmatrix} \cos\alpha & -\sin\alpha & 0 \\ \sin\alpha & \cos\alpha & 0 \\ 0 & 0 & 1 \end{bmatrix}$$

**Proof** By the lemma, there is an $e_3$, with $\|e_3\| = 1$, such that $Ae_3 = e_3$. By using the Gram-Schmidt process one obtains an orthogonal basis $e_1$, $e_2$, $e_3$ for $\mathbf{R}^3$. Clearly if V is the subspace spanned by $e_1$ and $e_2$, then V is invariant under A, that is $AV = V$ and $A|V$ is in SO(2). The matrix of $A|V$ with respect to $e_1$, $e_2$ is $\begin{bmatrix} \cos\alpha & -\sin\alpha \\ \sin\alpha & \cos\alpha \end{bmatrix}$.

In the proof of the next theorem we will explicitly need

**Corollary 2** If $A \in SO(3)$, $A \neq I$ then $\{x: Ax = x\}$ (the eigenspace with eigenvalue 1) is always 1-dimensional.

**Theorem 11** Let $G \subset SO(3)$ be a finite subgroup, then G is either cyclic, dihedral or the direct symmetry group of a Platonic solid.

**Proof** The group SO(3) preserves distances and preserves 0, hence it acts on

$$S^2 = \{x \in \mathbf{R}^3 \mid \|x\| = 1\}.$$

By Corollary 2 above, for each $g \neq 1$ in G, there is a unique set $\{x, -x\} \subset S^2$ such that $gx = x$ (and so $g(-x) = -x$), we call such an x a **pole** of g (x and $-x$ are the "north and south poles" of the rotation g).

Suppose $|G| = n$, and for each $x \in S^2$, let $m_x$ denote the number of elements in the subgroup

$$\text{Stab}(x) = \{g \in G \mid gx = x\},$$

that is, the number of elements (including the identity) of G that have x as a pole. Note that Stab(x) is a finite subgroup of SO(2) and so by Theorem 9 (and its proof) it is cyclic. The point $x \in S^2$ is a pole of some element $g \in G$ if and only if $m_x > 1$.

We will now study the orbit of a pole p under the action of G, that is

$$\mathrm{Orb}(p) = \{gp \mid g \in G\}.$$

It is a standard fact about groups acting on sets that $|\mathrm{Orb}(p)| = n/m_p$. The set Orb(p) consists entirely of poles because if $hp = p$ then $ghg^{-1}(gp) = gp$; and this also shows that $m_p = m_{gp}$, for each $g \in G$

As usual when studying orbits of actions of finite groups, one counts elements: for each pair of poles $\{p,-p\}$ we have $(m_p-1)$ elements $g \in G \setminus \{1\}$ for which they are poles, that is $gp = p$. For each orbit of poles we therefore have $n(m_p-1)/m_p$ elements of $G \setminus \{1\}$, but each element of $G \setminus \{1\}$ determines two poles p and $-p$, so the number of elements in $G \setminus \{1\}$ is

$$\tfrac{1}{2} \sum_{\text{orbits}} n(m_p-1)/m_p$$

where we have one summand for each orbit. Therefore one has the equation

$$2(n-1) = \sum_{\text{orbits}} n(m_p-1)/m_p$$

$$\text{or} \qquad 2(1-1/n) = \sum_{\text{orbits}} (1-1/m_p) \text{———————} (*)$$

To avoid the trivial group, one assumes that $n \geqslant 2$. By definition of a pole, $m_p \geqslant 2$. For each pole p one has $1 > (1-1/m_p) \geqslant \tfrac{1}{2}$. Also $2(1-1/n) \geqslant 1$. From this it is easy to deduce that there are either 2 or 3 orbits and we treat these cases separately.

**2 orbits:** Let the orbits have $n/m_1$ and $n/m_2$ elements. Equation (*) gives $2/n = 1/m_1 + 1/m_2$ and it follows that $n = m_1 = m_2$. In this case, therefore there are two orbits with exactly one pole in each. These must be fixed under every element of G and so one sees that G is cyclic, in fact this case is essentially two dimensional.

**3 orbits:** This is the full 3-dimensional situation. Equation (*) becomes

$$1 + 2/n = 1/m_1 + 1/m_2 + 1/m_3 \text{———————} (*_1)$$

and we assume $m_1 \geqslant m_2 \geqslant m_3$. It is clearly impossible to have $m_i \geqslant 3$ for each i for then the right hand side of $(*_1)$ could not exceed 1. So one has $m_3 = 2$. One therefore gets

$$1/2 + 2/n = 1/m_1 + 1/m_2 \text{———————} (*_2).$$

Similarly, it is impossible to have $m_1 \geqslant m_2 \geqslant 4$, so $m_2 = 2$ or 3. It is simpler to consider these two cases separately.

$\mathbf{m_2 = 2}$ In this case equation $(*_2)$ becomes $n = 2m_1$. Thus two of the orbits have $m_p = 2$ and have $m_1$ poles in them whilst the third orbit has $m_p = m_1$ and has 2 poles in it. An easy analysis shows that this is the dihedral case, the $2m_1$ poles being perpendicular to the other two poles. The $2m_1 = n$ poles are arranged in orbits, illustrated here for $n = 6$ and 8.

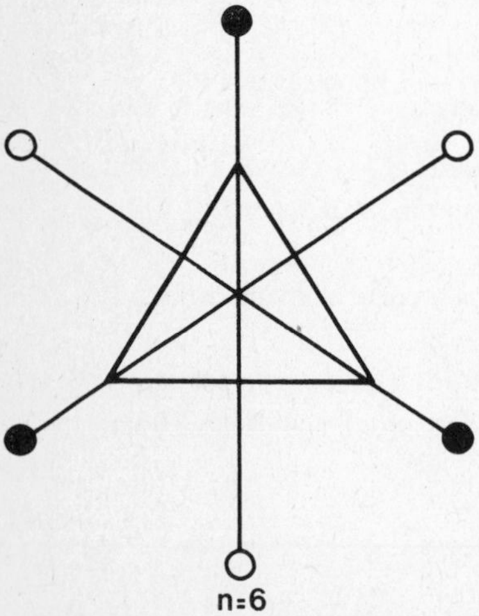

n=6

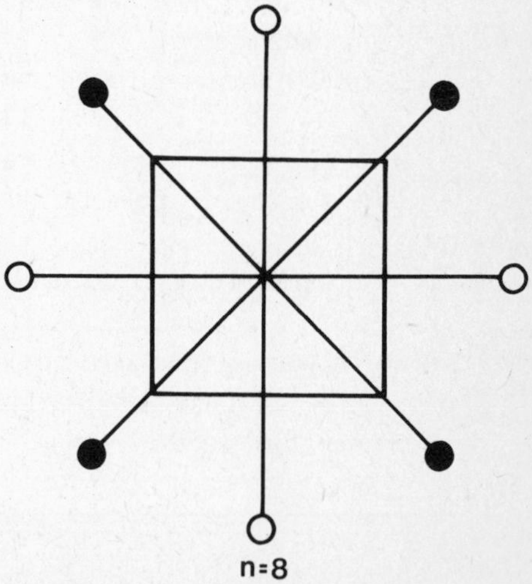

n=8

**$m_2 = 3$** In this case equation $(*_2)$ becomes

$$1/6 + 2/n = 1/m_1.$$

Clearly one has $3 \leqslant m_1 < 6$ and $(6-m_1)n = 12m_1$. The only solutions are

$$(m_1,n) = (3,12), (4,24) \text{ and } (5,60).$$

These numbers correspond to the Platonic solids. A full analysis (which we omit) shows that each of these cases can only arise from the symmetry group of a Platonic solid. The various orbits of poles correspond to the centres of faces, to the mid points of edges and to the vertices of the Platonic solid.

**Note** There is a difference between the dihedral group in $\mathbf{R}^2$ and the dihedral group of rotations in $\mathbf{R}^3$ although the groups are abstractly isomorphic.

An interesting application of these kinds of ideas is to the study of crystals.

## Crystals

In the mathematical theory of crystals it is assumed that the crystals are of infinite extent, an important concept in their theory is therefore that of a lattice in $\mathbf{R}^n$ (of course the main physical interest is in the case $n = 3$).

A **lattice** L in $\mathbf{R}^n$ is a subgroup of $\mathbf{R}^n$ that is discrete as a topological subspace of $\mathbf{R}^n$, and whose elements span $\mathbf{R}^n$. The discrete condition means that for each $x \in L$ there is an $\varepsilon > 0$ such that $d(x,y) \geqslant \varepsilon$ for all $y \in L$ with $y \neq x$, or equivalently that L has no accummulation point.

The following exercise should help to make this concept familiar.

**Exercise** i) Show that any line through two points of a lattice L contains infinitely many points of L.

ii) Show that L is isomorphic as a group to the subgroup of translations in its symmetry group S(L).

iii) If L is a lattice in $\mathbf{C}$ ($\cong \mathbf{R}^2$) such that $L = \bar{L}$ (L is self conjugate), prove that there is a basis for L whose vectors are two edges either of a rectangle whose edges are vertical and horizontal or of a rhombus whose diagonals are vertical and horizontal.

An alternative definition of a lattice $L \subset \mathbf{R}^n$ is that it is the set of integer combinations of a basis of $\mathbf{R}^n$,

$$L = \{\sum_{i=1}^{n} r_i e_i \mid r_i \in \mathbf{Z}\}.$$

Therefore L is isomorphic, as a group, to $\mathbf{Z}^n$. This definition is technically easier to use in proofs but, in practice, it is often easier to check that something is a discrete subgroup.

Given a discrete subgroup L of $\mathbf{R}^n$, one finds a basis for it as follows. Choose a non-zero vector $e_1$ in L that is closest to the origin. Having chosen $e_1, e_2, \ldots, e_k$ choose $e_{k+1}$ to be a vector in L not in $\text{Span}\{e_1, e_2, \ldots, e_k\}$ but closest to it. Some checking has to be done to see that L is the set of integer combinations of the vectors $e_1, e_2, \ldots, e_n$. The other main thing to check is that there is a closest vector at each stage.

An important restriction on the type of elements that can be in the symmetry group of a lattice is given by the following result – often called the '**crystallographic restriction**'.

**Theorem 12** If $L \subset \mathbf{R}^n$ is a lattice with $n = 2$ or 3 and $R \in S(L)$ is a rotation of order m, then $m = 2, 3, 4$ or 6.

**Proof** We give three proofs. The first, for $n = 2$, is simple. The result for $n = 3$ can be deduced from it by applying the fact (Corollary 1 on page 35) that any rotation of $\mathbf{R}^3$ is essentially a rotation of $\mathbf{R}^2$. The other two proofs are only sketched and are of importance because their ideas can be used to give corresponding results for dimensions $n > 3$.

1. Consider all the rotations in S(L). Their centres form a set $M \subset \mathbf{R}^2$. The set M contains L because the rotation through $\pi$ about any lattice point is in S(L). The set M is discrete: Let p be the centre of a rotation R in S(L) and $p \notin L$ and let q be a point of L closest to p, then p is the centroid of the orbit of q under the cyclic group $\{R\}$. Therefore p is not arbitrarily close to elements of L, hence M is discrete.

Let $R(p,2\pi/m)$ be the given rotation. Suppose $p_1$ is a point such that $R(p_1, 2\pi/m) \in S(L)$ and such that $d(p,p_1)$ is as small as possible. If $m \neq 2, 3, 4, 6$ we will show that there must be such a point closer to p than $p_1$. If $S \in S(L)$, then $SR(p,2\pi/m)S^{-1} = R(Sp,2\pi/m) \in S(L)$, so if $p_2 = R(p_1,2\pi/m)p$ we have that $p_2$ is the centre of a rotation through $2\pi/m$ which is a symmetry of L. But $d(p_2,p) = 2\sin(\pi/m)d(p,p_1)$ and so $d(p,p_2) < d(p,p_1)$ if $m > 6$.

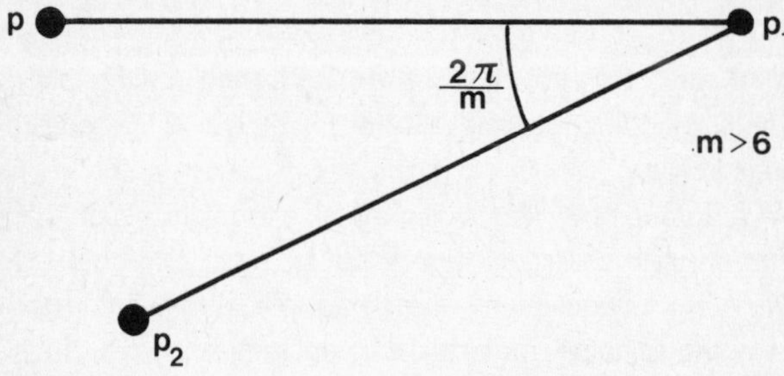


A similar argument shows that $p_3 = R(p_2,2\pi/m)p_1$ is such that $R(p_3,2\pi/m) \in S(L)$. If $m = 5$ a check shows that $d(p,p_3) < d(p,p_1)$ and so one concludes that $m \leqslant 6$ and $m \neq 5$.

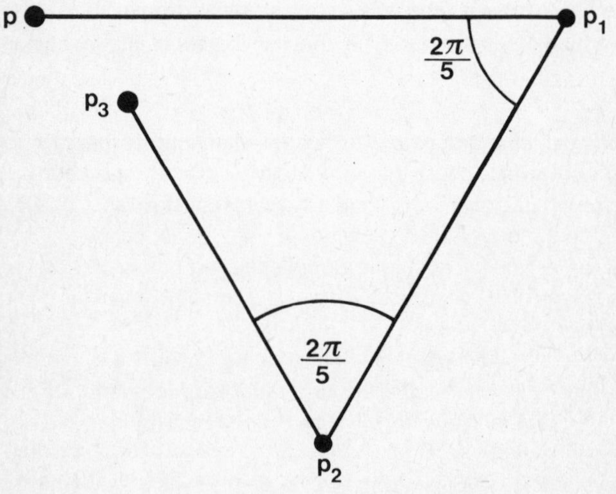

Each of these four remaining cases can occur, as seen by considering one of the following two lattices.

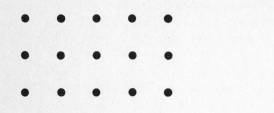

Square lattice $m = 2, 4$

Equilateral triangle lattice $m = 3,6$

2. If $f \in S(L)$ and f is linear, let $e_1, e_2, \ldots, e_n \in L$ be a basis for L so that L is the set of integral linear combinations of $e_1, e_2, \ldots, e_n$, then the matrix of f with respect to this basis has integer entries.

When $n = 2$ or 3 and f is a rotation there is an orthonormal basis for $\mathbf{R}^n$ with respect to which f is represented by the matrix

$$A_\theta = \begin{bmatrix} \cos\theta & -\sin\theta \\ \sin\theta & \cos\theta \end{bmatrix} \quad \text{or by} \quad \begin{bmatrix} A_\theta & 0 \\ 0 & 1 \end{bmatrix}$$

Under our hypotheses this matrix is similar to an integer matrix. As the trace of a matrix is invariant under changes of bases it must therefore be an integer. In either case we see that $2\cos\theta \in \mathbf{Z}$ and the only possibilities are $\cos\theta = 0, \pm 1/2, \pm 1$ giving that the rotation must have order 2, 3, 4 or 6.

As already mentioned this proof will give information in higher dimensions. For example when $n = 4$ the matrix is integral for one basis and is of the form $\begin{bmatrix} A_{\theta_1} & 0 \\ 0 & A_{\theta_2} \end{bmatrix}$ for another . So for $n = 4$ (and also for $n = 5$) one has $2(\cos\theta_1 + \cos\theta_2) \in \mathbf{Z}$. This approach seems to lead to difficult calculations in general and a slightly different approach is most tractable:

3. If $f \in S(L)$ is linear and of order m, then its minimal polynomial has integer coefficients and is of degree $\leqslant n$ and it also divides $x^m - 1$. These facts give information about m, for example:

**Exercise** i) If f is an irreducible transformation, show that $n \leqslant \varphi(m)$ where $\varphi$ is Euler's function.

ii) If $f \in S(L)$ is not linear (for example, f may be a rotation whose centre is not even a lattice point), explain carefully how to make proofs 2 and 3 complete.

It is clearly an interesting and important problem to try to describe all possible **crystal groups**, that is subgroups $G \subset \mathbf{I}(\mathbf{R}^n)$ which contain n independent translations but no arbitarily small translation. Before this can be done a definition of equivalence of two crystal groups must be agreed upon. This is a slightly thorny question and anyway the classification of the crystal groups even for $n = 2$ is too long to be included here. (A proof that there are exactly 17 crystal groups in dimension 2 is given in the article 'The seventeen plane symmetry groups' by R.L.E. Schwarzenberger in the Mathematical Gazette, vol. 58 (1974) pages 123-131. Readers of these notes should be able to read this article easily.) We will however give one classification result as Theorem 13.

The group $\mathbf{I}(\mathbf{R}^n)$ consisting of all isometries of $\mathbf{R}^n$ contains the translations $\mathbf{R}^n$ as a normal subgroup, and the quotient group $\mathbf{I}(\mathbf{R}^n)/\mathbf{R}^n$ is isomorphic to O(n). In fact, as a group, $\mathbf{I}(\mathbf{R}^n)$ can be described as the semi-direct product $O(n) \tilde{\times} \mathbf{R}^n$, that is, its elements are pairs (A,x) with $A \in O(n)$ and $x \in \mathbf{R}^n$ and the group operations are given by

$$(A,x).(B,y) = (AB, x+Ay)$$

and

$$(A,x)^{-1} = (A^{-1}, -A^{-1}x).$$

The action on $\mathbf{R}^n$ is given by $(A,x)v = x + Av$.

**Exercise** Let G be the group $\{1,a\}$ of order 2. It acts on $\mathbf{Z}$ by $a.n = -n$. Show that the semi-direct product $G \tilde{\times} \mathbf{Z}$ is the infinite dihedral group. Find a subset of $\mathbf{R}^2$ whose symmetry group is $G \tilde{\times} \mathbf{Z}$.

A crystal group $G \subset \mathbf{I}(\mathbf{R}^n)$ has a normal subgroup of translations, $G_t = G \cap \mathbf{R}^n$, that forms a lattice in $\mathbf{R}^n$. The quotient group $\bar{G} = G/G_t$ is called the **point group** of G. It is a subgroup of O(n) and is finite. Notice however that the point group may not be isomorphic to the stabilizer of any point in $\mathbf{R}^n$:

**Example** If $n = 2$, let G be the crystal group generated by a glide g of length one along the x-axis and a translation f of length one along the y-axis, in symbols

$$g(x,y) = g(x+1,-y), \qquad f(x,y) = f(x,y+1).$$

The translation subgroup $G_t$ is generated by f and $g^2$ and the quotient $\bar{G}$ has two elements. The stabilizer of any point is however the identity.

**Exercise** In the above example, if $\bar{G} = \{1,a\}$, calculate how a acts on f and on $g^2$. Show that $\bar{G} \tilde{\times} G_t$ is generated by f, $g^2$ and a where $a(x,y) = (x,-y)$. Prove that a does not lie in G and that g does not lie in $\bar{G} \tilde{\times} G_t$. Both G and $\bar{G} \tilde{\times} G_t$ contain $G_t$ as a subgroup of index 2.

**Lemma** If G is a crystal group, $\bar{G}$ acts on $G_t$.

**Proof** Given $A \in \bar{G}$, there is an $x \in \mathbf{R}^n$ such that $(A,x) \in G$. If $v \in G_t$ then $(I,v) \in G$. We show that $Av \in G_t$, that is $(I,Av) \in G$:

$$(A,x)(I,v)(A,x)^{-1} = (I,Av).$$

As this is independent of x, the action is well defined.

This lemma shows that the group $\bar{G}$ acts on the lattice $G_t$ by orthogonal transformations. Hence the elements of $\bar{G}$ satisfy the crystallographic restriction even though the group $\bar{G}$ may not be a subgroup of G. We will now proceed to say something about the classification of the various point groups that arise in dimension 3. A point group is a finite subgroup of O(3) and satisfies the crystallographic restriction. We will classify such groups up to conjugation in O(3), that is, we will say that two subgroups $G_1$, $G_2$ are equivalent if they are conjugate. Two subgroups of O(3) that are isomorphic may not be conjugate, they are conjugate only if they act on $\mathbf{R}^3$ in the same way.

**Exercise** Show that any element of order 2 in O(3) is conjugate to one of the three elements

$$\begin{bmatrix} -1 & 0 & 0 \\ 0 & -1 & 0 \\ 0 & 0 & -1 \end{bmatrix} \quad \begin{bmatrix} -1 & 0 & 0 \\ 0 & -1 & 0 \\ 0 & 0 & 1 \end{bmatrix} \quad \begin{bmatrix} -1 & 0 & 0 \\ 0 & 1 & 0 \\ 0 & 0 & 1 \end{bmatrix}$$

and that these are not conjugate to each other.

Now consider a finite subgroup $\bar{G} \subset O(3)$ all of whose elements have order 2, 3, 4 or 6. If $\bar{G}$ lies in SO(3) then by Theorem 11 it must be one of the following list.

$$1, \mathbf{Z}/2, \mathbf{Z}/3, \mathbf{Z}/4, \mathbf{Z}/6, D_2, D_3, D_4, D_6, \text{Tet}, \text{Oct} \qquad (*)$$

where Tet($\cong A_4$) and Oct($\cong S_4$) denote the rotation groups of the tetrahedron and octahedron respectively. The classification given in Theorem 11 was also up to conjugacy.

Now suppose $\bar{G}$ is not entirely contained in SO(3) so that $\bar{G}$ has a normal subgroup $\bar{G}_d$ of index 2 consisting of the direct elements. The group $\bar{G}_d$ is in the list (*). Then $\bar{G} = \bar{G}_d \cap g\bar{G}_d$ for any $g \in \bar{G} \setminus \bar{G}_d$. If

$$J = \begin{bmatrix} -1 & 0 & 0 \\ 0 & -1 & 0 \\ 0 & 0 & -1 \end{bmatrix}$$

is in $\bar{G}$ then $\bar{G} = \bar{G}_d \times \{1,J\}$ (compare the argument on page 29). This possibility gives a further eleven groups. There remains the case that $J \notin \bar{G}$, so $\bar{G} = \bar{G}_d \cup JX$ for a subset X of SO(3). Because $\bar{G}$ is a group and J commutes with every element in O(3), the set $Y = \bar{G}_d \cup X$ is a subgroup of SO(3), and $|\bar{G}| = |Y|$. Every lattice has J as a symmetry, so Y preserves the lattice that $\bar{G}$ preserves. Hence Y is in the list (*), and has a normal subgroup of index 2. An examination of the various possibilities shows that a further ten groups $\bar{G}$ arise in this way, so we obtain

**Theorem 13** Up to conjugacy there are exactly 32 finite subgroups of O(3) that satisfy the crystallographic restriction.

Each of these groups arises as the point group of some crystal group, and with one exception they all occur as a point group of some known, physically occurring crystal. In dimension three the total number of crystal groups is 230. It was proved in 1910 by Bieberbach that there are only a finite number of crystal groups in each dimension.

For more information the reader is urged to consult books on crystallography; a good first exercise would be to search in the book for the equivalent of Theorem 13. A suitable introduction is the book 'Elementary Crystallography' by M.J. Buerger (published by J. Wiley in 1956).

It is interesting to note that this discussion is quite close to an unsolved mathematical problem:

How densely can one pack balls of equal radius in $\mathbf{R}^3$?

Let us assume that the balls have radius 1 and let $I_\ell$ be the cube with edges of length $2\ell$ and centre at 0. The density of a packing of $\mathbf{R}^3$ by an arrangement of non overlapping balls is defined to be

$$\lim_{\ell \to \infty} \left( \frac{\text{vol of balls in } I_\ell}{8\ell^3} \right)$$

For any packing this limit is known to be greater than a number that is approximately 0.7796. A lattice gives rise to a packing by putting the centres of the balls at the lattice points and then changing the radius of the balls until they just touch. The densest lattice packing comes from the 'face centred cubic' lattice, its density is $\pi/\sqrt{18} \simeq 0.7405$, and it is conjectured that this is the densest possible packing of $\mathbf{R}^3$.

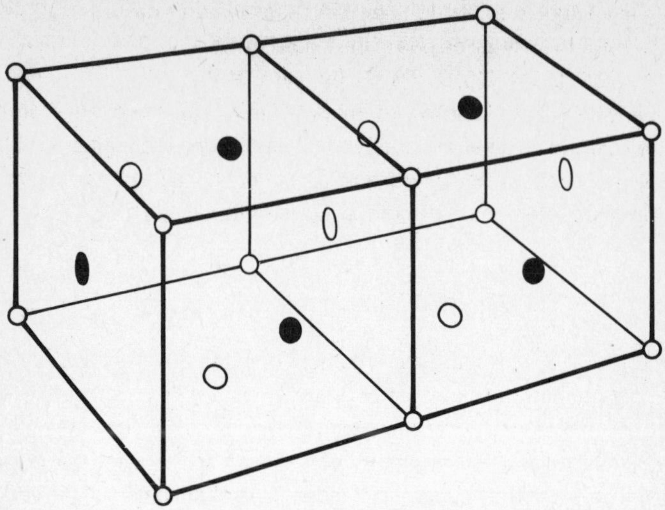

face centred cubic lattice

The face centred cubic lattice has points at the vertices of a family of cubes and at the centre of each face. It occurs in the crystal structure of gold.

**Exercise** i) Prove that the face centred cubic lattice is spanned by the (½,½,0), (½,0,½) and (0,½,½).

ii) Find the densest packing of $\mathbf{R}^2$.

**Rotations and Quaternions**

There is an important connection between the group of rotations SO(3) and the quaternions. Before describing this connection we will study SO(3) as a topological space. The 3-dimensional ball (or disc) of radius r is denoted by

$$D_r^3 = \{x \in \mathbf{R}^3 \mid \|x\| \leqslant r\}.$$

**Theorem 14** The space SO(3) is homeomorphic to the quotient space $D_1^3/\sim$ where $\sim$ is the equivalence relation

$$x \sim y \text{ if and only if } \qquad x = y \qquad \text{or} \qquad x = -y \text{ and } \|x\| = 1.$$

(If you need reminding about how to make a quotient space into a topological space, look again at the discussion about the Möbius band on page 18.)

**Proof** By Corollary 1 on page 35 each element of SO(3) is determined by its axis and the angle of rotation. Consider the disc $D^3_\pi$ of radius $\pi$. Define a map $f: D^3_\pi \to SO(3)$ as follows:

f(x) is the rotation through angle $\|x\|$ about the axis Ox.

Note that f(x) is well defined, for f(0) is the identity, so it does not matter that for x = 0 one does not have an axis. By Corollary 1 on page 35, f maps $D^3_\pi$ onto SO(3). Moreover f is injective except that when $\|x\| = \pi$ one has f(x) = f(−x). Hence it follows that f defines a continuous bijection between $D^3_\pi/\sim$ and SO(3); as both of these spaces are compact f induces a homeomorphism between them.

The **quaternions** are defined in a somewhat similar way to the complex numbers. The complex numbers **C** are defined as a vector space over **R** with basis $\{1,i\}$ and multiplication is defined to be bilinear, $i^2 = -1$ and elements of **R** commute with everything. Similarly the quaternions **H** (**H** stands for Hamilton their discoverer; one cannot use **Q** as that is reserved for the rational numbers) form a four dimensional vector space over **R** with basis $\{1,i,j,k\}$, the multiplication is again bilinear and **R** is central, also

$$i^2 = j^2 = k^2 = -1$$
$$ij = k,\ jk = i,\ ki = j,\ ji = -k,\ kj = -i,\ ik = -j$$

(for these last relations use + sign if the symbols are in the cyclic order i, j, k otherwise use − sign).

**Exercise** Verify that $(x+iy+jz+ku)(x'+iy'+jz'+ku')$ equals

$$xx'-yy'-zz'-uu' +i(xy'+yx'+zu'-uz') + j(xz'-yu'+zx'+uy') + k(xu'+yz'-zy'+ux').$$

An alternative approach is to consider the subset of $M(2,\mathbf{C})$ consisting of all the matrices of the form $\begin{bmatrix} \alpha & \beta \\ -\bar\beta & \bar\alpha \end{bmatrix}$. The quaternion $x + iy + jz + ku$ is represented by the matrix

$$\begin{bmatrix} x+iy & z+iu \\ -z+iu & x-iy \end{bmatrix}$$

Matrix multiplication corresponds to quaternion multiplication. The quaternion q can be regarded as a point in $\mathbf{R}^4$, and so has a norm $\|q\|$. If q is represented by $A = \begin{bmatrix} \alpha & \beta \\ -\bar\beta & \bar\alpha \end{bmatrix}$ then $\|q\|^2 = \det A$ and so $\|q_1q_2\| = \|q_1\|\|q_2\|$.

A quaternion q has a **conjugate** $\bar q = x - iy - jz - ku$. Then $q\bar q = \|q\|^2$, and hence a non-zero quaternion q has an **inverse** $q^{-1} = \bar q / \|q\|^2$. Thus **H** is a non-commutative division ring, and in many respects behaves like a field. It is now easy to verify that the 3-dimensional sphere

$$S^3 = \{q \in \mathbf{H} \mid \|q\| = 1\}$$

is a group – the **group of unit quaternions**.

Let U(n) denote the **unitary group**

$$U(n) = \{A \in M(n,\mathbf{C}) | A\bar A^t = I\}$$

and SU(n) denotes the **special unitary group**, those unitary matrices with determinant 1. One easily checks that the above correspondence between $2 \times 2$ matrices and quaternions gives rise to an isomorphism between the groups SU(2) and $S^3$. The rows of a unitary matrix are orthonormal with respect to the Hermitian inner product

$$<x,y> = \sum_{i=1}^{n} x_i\bar y_i \text{ on } \mathbf{C}^n.$$

Hence an element of U(2) is of the form $A = \begin{bmatrix} \alpha & \beta \\ -\lambda\bar\beta & \lambda\bar\alpha \end{bmatrix}$ with $\alpha\bar\alpha + \beta\bar\beta = 1$ and $|\lambda| = 1$. The matrix A lies in SU(2) if and only if $\lambda = 1$, and the inverse of such an A is $\begin{bmatrix} \bar\alpha & -\beta \\ \bar\beta & \alpha \end{bmatrix}$.

To make the connection between the unit quaternions and rotations one needs a three dimensional real vector space on which $S^3$ acts. This is the space $\mathbf{R}^3 = \{iy+jz+ku\}$ of purely imaginary quaternions.

**Proposition** The group $S^3$ acts orthogonally on $\mathbf{R}^3$ by conjugation.

**Proof** We make use of the representation of quaternions as $2 \times 2$ complex matrices of the form $\begin{bmatrix} \alpha & \beta \\ -\bar\beta & \bar\alpha \end{bmatrix}$. Such a matrix represents a purely imaginary quaternion if and only if its trace vanishes. As the trace of a matrix is preserved by conjugation, that is,

$$\operatorname{tr}(PAP^{-1}) = \operatorname{tr}(A),$$

one sees that if $q \in S^3$ and $q_1 \in \mathbf{R}^3$ then $qq_1q^{-1}$ is also in $\mathbf{R}^3$. So for each $q \in S^3$ one has a map $\rho(q)\colon R^3 \to R^3$ defined by $\rho(q)q_1 = qq_1q^{-1}$. It is clear that the map $\rho(q)$ is linear; it is also norm-preserving because

$$\|qq_1q^{-1}\| = \|q\|.\|q_1\|.\|q\|^{-1} = \|q_1\|.$$

So by the lemma on page 4, $\rho(q)$ is orthogonal.

**Theorem 15** Conjugation of quaternions induces an isomorphism $\rho: S^3/\{\pm 1\} \to SO(3)$.

**Proof** The proposition shows that $\rho: S^3 \to O(3)$ is a homomorphism. The space $S^3$ is connected and $\rho$ is continuous so the image of $\rho$ is connected. The group O(3) has two components, the component of the identity is SO(3) and so the image of $\rho$ is in SO(3). If $\rho(q) = I$ then $qq_1q^{-1} = q_1$ for all $q_1 \in \mathbf{R}^3$. A straightforward check shows that q is real. The only reals in $S^3$ are $\pm 1$, so the kernel of $\rho$ is $\{\pm 1\}$. It remains to prove that $\rho$ is onto. Consider $q = \lambda + a\mu$ with $a \in \mathbf{R}^3$, $\|a\| = 1$, $\lambda, \mu \in \mathbf{R}$ and $\lambda^2 + \mu^2 = 1$. Clearly $qaq^{-1} = a$ so that $\rho(q)$ is a rotation with axis a. The trace of a rotation with angle $\theta$ is $1 + 2\cos\theta$ so to determine the angle of rotation we calculate $\mathrm{tr}\rho(q)$. The diagonal entries of the matrix of $\rho(q)$ with respect to the basis i, j, k are Re(–iqiq), Re(–jqjq) and Re(–kqkq). If $a = a_1 i + a_2 j + a_3 k$ and $q = \lambda + a\mu$ an easy calculation shows that $\mathrm{Re}(-iqiq) = \lambda^2 + \mu^2(a_1{}^2 - a_2{}^2 - a_3{}^2)$. Similarly for j and k. So

$$\mathrm{tr}(\rho(q)) = 3\lambda^2 - \mu^2\|a\|^2 = 3\lambda^2 - \mu^2 = 1 + 2(\lambda^2 - \mu^2) = 1 + 2\cos\theta$$

if $\lambda = \cos\theta/2$, $\mu = \sin\theta/2$.

Hence, by choosing $\lambda$, $\mu$ and a suitably, one can write any rotation in the form $\rho(q)$.

From this theorem it follows that SO(3) is homeomorphic to the space $S^3/x \sim -x$. This fact ties in well with Theorem 14: the lower hemisphere of $S^3$ given by $u \leqslant 0$ is clearly homeomorphic to $D^3$ by the projection $(x,y,z,u) \to (x,y,z)$. Every point of $S^3/x \sim -x$ has a representative in the lower hemisphere, and this representative is unique unless $u = 0$. A point of $S^3/x \sim -x$ with $u = 0$ has two representatives in the lower hemisphere and they are antipodal points of the boundary of $D^3$, that is, points $x, -x$ with $\|x\| = 1$. This discussion gives an alternative proof of Theorem 14.

The description of the group SO(3) as $S^3/\{\pm 1\}$, or equivalently as $SU(2)/\{\pm I\}$, is useful in quantum mechanics where it is used for the study of spin.

# Problems

1. If f is a distance preserving map of $\mathbf{R}^n$ to itself, show that f is onto. Try to find conditions on a metric space X that ensure that any distance preserving map $f: X \to X$ is onto.
2. Show that $GL(n,\mathbf{C})$ is homeomorphic to $U(n) \times \mathbf{R}^{n^2}$ where U(n) is the space of unitary matrices.
3. Describe the space of all real $2 \times 2$ matrices of rank 1.
4. Show that $M(n,\mathbf{R})$ has a linear subspace V of dimension n(n–1) which consists entirely of singular matrices. If $n > 1$, show that there are infinitely many such spaces. It is true that there is no linear space V of singular matrices in $M(n,\mathbf{R})$ with $\dim V > n(n-1)$, but this is much harder to prove.
5. Show that $\mathbf{I}_d(\mathbf{R}^2)$ is isomorphic to the group of all complex $2 \times 2$ matrices of the form $\begin{bmatrix} w & z \\ 0 & 1 \end{bmatrix}$ with $|w| = 1$.
6. Find a group of $(n+1) \times (n+1)$ real matrices that is isomorphic to $\mathbf{I}(\mathbf{R}^n)$.
7. Let a, b be unit vectors in $\mathbf{R}^2$ and R be the (unique) rotation about the origin such that $Ra = b$.
   i) Express Rb in terms of a, b and a.b.
   ii) Express Rx (for any $x \in \mathbf{R}^2$) in terms of a, b, x, a.b, a.x and b.x.
8. Work out the compositions of the various elements in $\mathbf{I}(\mathbf{R}^2)$; for example, what is $R(a,\alpha)R(b,-\alpha)$ for $a \neq b$?
9. Suppose that $f: \mathbf{R}^n \to \mathbf{R}^n$ is an isometry such that there is a $K > 0$ with $d(fx,x) < K$ for all $x \in \mathbf{R}^n$, show that f is a translation.
10. If $A \in O(n)$, show that there is an orthonormal basis for $\mathbf{R}^n$ with respect to which the matrix of A has elements of O(1) and O(2) along the diagonal. Deduce that any $f \in \mathbf{I}(\mathbf{R}^n)$ can be written in the form
$$T^aH^bR^c \text{ with } a, b = 0 \text{ or } 1, c \geqslant 0 \text{ and } a + b + 2c \leqslant n,$$
where T is a translation, H is a reflection and $R^c$ is the composition of c rotations about (n–2) dimensional subspaces. Show further that different triples (a,b,c) always give different types of isometries. Hence show that, apart from the identity, there are 2n types of isometries of $\mathbf{R}^n$.
11. The **half turn** $H_a$ is the rotation through $\pi$ about the point a in $\mathbf{R}^2$. Show that
    i) the product of two half turns is a translation,
    ii) every translation can be written as $H_aH_b$ and that either a or b can be chosen arbitrarily,
    iii) every opposite isometry is the product of a half turn and a reflection,
    iv) $H_aH_bH_c = H_cH_bH_a$ and is a half turn.
12. Show that $(R_\ell R_m R_n)^2$ is a translation and give necessary and sufficient conditions for $R_\ell R_m R_n$ to be a reflection. If $\ell$, m, n are the sides of a triangle, find the compositions of $R_\ell$, $R_m$, $R_n$ in the various orders.

13. Prove that
    i) $R_\ell R_m = R_m R_\ell \Leftrightarrow$ m is perpendicular to $\ell$,
    ii) $H_a R_\ell = R_\ell H_a \Leftrightarrow a \in \ell$,
    iii) $R_\ell R_m R_n = R_n R_m R_\ell \Leftrightarrow \ell$, m, n are concurrent or parallel,
    iv) $H_a H_b = H_b H_c \Leftrightarrow$ b is the mid-point of ac,
    v) $H_a R_\ell = R_\ell H_b \Leftrightarrow \ell$ is the perpendicular bisector of ab.

    Translate other conditions about points and lines into ones about reflections and half turns.

14. i) If f and g are direct isometries of $\mathbf{R}^2$, prove that $fgf^{-1}g^{-1}$ is a translation. Show that every translation can be written in this way.
    ii) Show that every direct isometry of $\mathbf{R}^2$ can be written in the form $fgf^{-1}g^{-1}$, where f, g are isometries of $\mathbf{R}^2$.

15. Let a, b, c be a clockwise oriented triangle in $\mathbf{R}^2$ whose angles are $\alpha, \beta, \gamma$ at a, b, c respectively. Let $R(p,\theta)$ denote rotation about p through $\theta$ anticlockwise. Prove that
$$R(c,2\gamma)R(b,2\beta)R(a,2\alpha) \qquad \text{is the identity.}$$

16. If a, b, c are affinely independent points in $\mathbf{R}^2$ then $x \in \mathbf{R}^2$ can be written uniquely as $x = \lambda a + \mu b + \nu c$ with $\lambda + \mu + \nu = 1$. Prove that the ratios $\lambda{:}\mu{:}\nu$ are equal to the ratios $\triangle(xbc){:}\triangle(axc){:}\triangle(abx)$ where $\triangle(xbc)$ denotes the area of the triangle xbc; this area is to be interpreted with a certain sign which you should explain.

17. If a, b, c and $a^1, b^1, c^1$ are two sets of affinely independent points in $\mathbf{R}^2$, show that there is a unique map $f: \mathbf{R}^2 \to \mathbf{R}^2$ of the form $fx = Ax + d$ where $A \in GL(2,\mathbf{R})$ and $d \in \mathbf{R}^2$ such that $fa = a^1$, $fb = b^1$, $fc = c^1$. Show that
$$\triangle(a^1b^1c^1) = \det A . \triangle(abc)$$
where $\triangle$ denotes the area of a triangle.

    Let $a_ix + b_iy + c_i = 0$ $(i = 1, 2, 3)$ be three non-concurrent lines in $\mathbf{R}^2$ such that no pair is parallel. By using a suitable affine transformation, show that the area of the triangle bounded by the three lines is $(\det A)^2/2|\det C_1 \det C_2 \det C_3|$, where A is the matrix
$$\begin{bmatrix} a_1 & b_1 & c_1 \\ a_2 & b_2 & c_2 \\ a_3 & b_3 & c_3 \end{bmatrix}$$
and $C_i$ is the $2 \times 2$ cofactor matrix of the entry $c_i$.

18. Prove that any two reflections of $\mathbf{R}^n$ are conjugate in $\mathbf{I}(\mathbf{R}^n)$. If G is any abelian group and
$$\alpha: \mathbf{I}(\mathbf{R}^n) \to G$$
is a homomorphism, show that $\alpha(R_1) = \alpha(R_2)$ for any two reflections $R_1, R_2$. Deduce that there is essentially only one non-trivial homomorphism from $\mathbf{I}(\mathbf{R}^n)$ to any abelian group, namely the sign homomorphism defined on page 28.

19.* If H is a linear hyperplane in $\mathbf{R}^n$, let
$$\Gamma_H = \{f \in GL(n,\mathbf{R}) \mid fx = x \text{ for } x \in H\}.$$
If a is a point not in M, prove that there is a real number $\lambda$ independent of a such that $fa = \lambda a + h$ for some $h \in H$. If $\lambda = 1$, f is called a transvection. Show that the set of transvections is normal in $\Gamma_H$. If $\lambda \neq 1$, f is called a dilation. Prove that every element of $GL(n,\mathbf{R})$ can be written as a product of transvections and dilations, and investigate the extent to which this can be done uniquely.

20. Let Aff($\mathbf{R}^n$,$\mathbf{R}^n$) denote the set of all affine maps f: $\mathbf{R}^n \to \mathbf{R}^n$. Let $L_f$ denote the linear map that corresponds to f, that is $L_f(x) = f(x) - f(0)$. Show that

$$L: \mathrm{Aff}(\mathbf{R}^n,\mathbf{R}^n) \to M(n,\mathbf{R})$$

is a map of rings.

By identifying $\mathbf{R}^n$ with the set $\{x \in \mathbf{R}^{n+1} | x_{n+1} = 1\}$ show that Aff($\mathbf{R}^n$,$\mathbf{R}^n$) can be identified with a subset of M(n+1,$\mathbf{R}$). Multiplication corresponds under this identification but addition does not.

21.* An **affinity** of $\mathbf{R}^n$ is an affine isomorphism of $\mathbf{R}^n$.

i) Show that the set of all affinities of $\mathbf{R}^n$ form a group $\mathbf{A}(\mathbf{R}^n)$ and that the set of translations form a normal subgroup.

ii) When $\mathbf{A}(\mathbf{R}^n)$ is given its obvious topology, show that it is homeomorphic to GL(n,$\mathbf{R}$) $\times$ $\mathbf{R}^n$.

iii) If $\{a_0, a_1, \ldots, a_n\}$ and $\{b_0, b_1, \ldots, b_n\}$ are two sets of independent points in $\mathbf{R}^n$, prove that there is a unique affinity f such that $fa_i = b_i$ for each i. Use this fact to find the dimension of the space $\mathbf{A}(\mathbf{R}^n)$ and check that it agrees with that obtained from ii).

iv) If X is a subset of $\mathbf{R}^n$ whose centroid is at c, show that the centroid of fX is at fc.

v) Which of the following concepts are preserved by an affinity of $\mathbf{R}^2$? Circle, ellipse, hyperbola, parabola, parallelogram, mid-point of a line segment.

22.* If $\ell$ is a line in $\mathbf{R}^2$, let Fix($\ell$) = $\{g \in GL(2,\mathbf{R}) | gx=x, \forall x \in \ell\}$. Prove that Fix($\ell$) is homeomorphic to two copies of $\mathbf{R}^2$ and is isomorphic to the group $\mathbf{A}(\mathbf{R}^1)$. Generalise these facts to higher dimensions.

Describe Fix($\ell$) as a subspace of GL(2,$\mathbf{R}$), remembering that GL(2,$\mathbf{R}$) is O(2) $\times$ $\mathbf{R}^3$. (To visualize this, it may help to consider Fix($\ell$) $\cap$ SL(2,$\mathbf{R}$); SL(2,$\mathbf{R}$) is homeomorphic to an open torus.)

Let Inv($\ell$) = $\{g \in GL(2,\mathbf{R}) | gx \in \ell, \forall x \in \ell\}$. Show that Inv($\ell$) is homeomorphic to the union of four copies of $\mathbf{R}^3$, and investigate how it sits inside GL(2,$\mathbf{R}$).

23.* If we represent the circle C in $\mathbf{R}^2$ with centre (p,q) and radius R by the point $P_C = (p,q,r) \in \mathbf{R}^3$ where $r = p^2 + q^2 - R^2$, describe the subset X of $\mathbf{R}^3$ that corresponds to the set of all circles in $\mathbf{R}^2$.

Verify that the family of circles that corresponds to the points of a line in $\mathbf{R}^3$ is a coaxal family, and show that there are three types of coaxal families corresponding to the cases where the line meets, misses or touches the boundary of X.

If $\ell$, m are skew lines in $\mathbf{R}^3$ and $a \notin \ell \cup m$, prove that there is a line n such that $a \in n$, $\ell \cap n \neq \phi$ and $m \cap n \neq \phi$. If x,x′,y,y′ are four non-concyclic points in $\mathbf{R}^2$ one of them lying on the circle C, then there are points z, z′ $\in$ C such that {x,x′,z,z′} and {y,y′,z,z′} are concyclic.

24.* If $r^2 = x^2 + y^2$, then the subset of $\mathbf{R}^3$ defined by $(r-a)^2 + z^2 = b^2$ $(a > b)$ is a torus. The torus is the union of a family of circles, the **meridians**, all of radius b. They are the various images of the circle C: $(x-a)^2 + z^2 = b^2$, $y = 0$ under the various rotations about the z-axis. It is also the union of another (orthogonal) family of circles, the **parallels**, whose radii vary from $a - b$ to $a + b$. They are the circles of revolution obtained as the images of single points of the circle C as C revolves round the z-axis.

There are other interesting families of circles on the torus:
Show that each of the planes

$$\pi_\alpha : b(x \cos\alpha + y \sin\alpha) = \pm z \sqrt{(a^2-b^2)}$$

meets the sphere (whose radius is a)

$$S_\alpha : x^2 + y^2 + z^2 + 2b(x \sin\alpha - y \cos\alpha) = a^2 - b^2$$

in a great circle of radius a that lies on the torus. As $\alpha$ varies one gets two families of circles on the torus (corresponding to the $\pm$ in the equation of the plane). Verify that the circles of one system do not meet each other, and that they cover the torus, but that two circles belonging to different systems meet each other in exactly two points. (This example is treated in Coxeter, 'Introduction to Geometry' pages 132-133.)

25. Let G be a group generated by two elements a, b both of order two. If the order of ab is n show that G is isomorphic to $D_n$, the dihedral group of order n. (Dihedral groups are characterized by the property that they are generated by two elements of order two.)

26. Show that there are exactly seven distinct infinite subgroups $G \subset \mathbf{I}(\mathbf{R}^2)$ such that Z is an orbit of G, where $Z = \{(n,0): n \in \mathbf{Z}\}$. For each such G, draw a subset $X(G) \subset \mathbf{R}^2$ such that $S(X(G)) = G$. Sometimes these subsets in $\mathbf{R}^2$ are called the seven wallpaper patterns.

27. Draw a 'plan' projection of the dodecahedron in a plane perpendicular to a face. Do the same for an icosahedron in a plane perpendicular to a main diagonal.

28. Show that any two diagonals of a regular pentagon divide each other in the ratio $\tau$:1 – the 'golden section' $\tau$ equals $(1+\sqrt{5})/2$.

    The twelve vertices of a regular icosahedron of edge length 2 are the vertices of three mutually perpendicular (golden) rectangles whose sides are of length 2 and $2\tau$. Draw them. If these rectangles are in the three co-ordinate planes, verify that the twelve vertices have co-ordinates $(0,\pm\tau,\pm1)$, $(\pm1,0,\pm\tau)$ and $(\pm\tau,\pm1,0)$. Show that the octahedron whose six vertices are at the points $(\pm\tau^2,0,0)$, $(0,\pm\tau^2,0)$ and $(0,0,\pm\tau^2)$ is regular. The twelve vertices of the icosahedron lie on the twelve edges of the octahdron and divide them in the ratio $\tau$:1.

    (For more on $\tau$ see Coxeter 'Introduction to Geometry' chapter 11.)

29. Which of the Platonic solids have the property that their faces can be coloured black and white and so that every pair of adjacent faces are differently coloured?

30. Given one edge of the dodecahedron, show that there are exactly five others that are either parallel or perpendicular to it. In this way the thirty edges can be divided into five sets of six edges. Describe the precise relationship of these sets to the five inscribed cubes and show that this gives another way of understanding the symmetries of the dodecahedron. This method also works for the icosahedron.

31. If $\Gamma$ is a connected graph in $\mathbf{R}^2$ with V vertices, E edges and F faces, use induction on E to show that $V - E + F = 1$. (There will be two cases – the new edge may join two old vertices or it may have a new vertex. If this is not enough of a hint consult Coxeter 'Introduction to Geometry' page 152.)

    Consider a graph $\Gamma$ on $S^2$. By projecting $\Gamma$ stereographically to $\mathbf{R}^2$ (see page 82) from a point not on $\Gamma$ to obtain a graph $\Gamma_1 \subset \mathbf{R}^2$ and using the above result for $\Gamma_1$, prove that $V - E + F = 2$ for $\Gamma$.

    What is $V - E + F$ for a graph on a torus? Deduce the formula from the above result for a graph in $\mathbf{R}^2$.

32. A convex polyhedron has $f_n$ faces with n edges and $v_n$ vertices at which n edges meet. Show that
i) $\Sigma nf_n = \Sigma nv_n$,
ii) $\Sigma f_{2n+1}$ is even,
iii) $v_3 + f_3 > 0$.

33. If L is a lattice in $\mathbf{R}^n$, show that there is a compact subset $F \subset \mathbf{R}^n$ such that the sets $x + F$ $(x \in L)$ cover $\mathbf{R}^n$ and the sets $x + \mathring{F}$ are all disjoint.

34.* A basis $\{e_1, e_2, \ldots, e_n\}$ for $\mathbf{R}^n$ determines a parallelepiped
$$P = \{\underline{x} \in \mathbf{R}^n \mid \underline{x} = \sum_{i=1}^{n} x_i e_i \text{ with } 0 \leqslant x_i \leqslant 1\}$$
and a lattice
$$L = \{\underline{x} \in \mathbf{R}^n \mid \underline{x} = \sum_{i=1}^{n} m_i e_i \text{ with } m_i \in \mathbf{Z}\}.$$
Let $S = \max\{\|e_1\|, \|e_2\|, \ldots, \|e_n\|\}$ be the length of the longest edge of P. Prove that, for $n \leqslant 4$, if $x \in P$ then there is an $e \in L$ such that $\|x-e\| \leqslant S$. Show that this is false for $n \geqslant 5$.

35.* i) Show that any element $A \in SU(2)$ is conjugate to a diagonal matrix $U_\varphi = \begin{bmatrix} e^{i\varphi} & 0 \\ 0 & e^{-i\varphi} \end{bmatrix}$. That is, there is a $P \in SU(2)$ such that $P^{-1}AP = U_\varphi$. The angle $\varphi$ can be chosen uniquely in $[0, \pi]$ and is $\cos^{-1}(\mathrm{tr}A/2)$.
ii) If $B = \begin{bmatrix} \alpha & -\beta \\ \bar{\beta} & \bar{\alpha} \end{bmatrix}$, show that $\mathrm{tr}(BU_\varphi B^{-1} U_\varphi^{-1}) = 2(\alpha\bar{\alpha} + \beta\bar{\beta}\cos 2\varphi)$.
iii) Let H be a normal subgroup of SU(2) that contains an element A such that $\mathrm{tr}A = 2\cos\varphi_0 \neq \pm 2$. Use i) and ii) to show that H contains matrices with all possible traces between 2 and $2\cos 2\varphi_0$. Hence show that $H = SU(2)$.
iv) Deduce that if $H_1 \subset SO(3)$ is a normal subgroup then $H_1 = \{1\}$ or SO(3). This means that SO(3) is a **simple** group.

36. Prove that two non-zero quaternions x, y are orthogonal if and only if $x^{-1}y$ is purely imaginary.

37. Regard a purely imaginary quaternion as a vector in $\mathbf{R}^3$ and vice versa. Show that the quaternion product xy of $x, y \in \mathbf{R}^3$ is $-x.y + x_x y$ ($x_x y$ is the vector product of x and y).
If I, J, K is a right-handed set of orthonormal vectors in $\mathbf{R}^3$, show that $I^2 = J^2 = K^2 = -1$ and
$$IJ = K, JK = I, KI = J, JI = -K, KJ = -I, IK = -J.$$

38. If $q \in S^3$, write $q = \cos\theta + I\sin\theta$ with $I \in S^2$. Prove that $\rho(q) \in SO(3)$ is the rotation through $2\theta$ about the axis OI.

39. Show that the elements $\pm 1, \pm i, \pm j, \pm k$ and $(\pm 1 \pm i \pm j \pm k)/2$ form a subgroup of $S^3$. What is its image in SO(3)?

40. (For those with a knowledge of the calculus of several variables.)
The spaces $S^3$ and SO(3) are both connected, compact, closed, smooth manifolds of dimension three. The map $\rho: S^3 \to SO(3)$ is smooth, everywhere regular and $|\rho^{-1}x| = 2$ for each $x \in SO(3)$. Deduce that $\rho$ is onto.

# Part II

# Projective Geometry

Projective geometry was invented in the 17th century, the first important contributions to the subject being made by a French architect Gérard Desargues. These studies arose from attempts to understand the geometrical properties of perspective drawing. When one draws a pair of parallel lines, such as the sides of a long, straight road, it is usual to draw them as meeting 'at infinity'. In the drawing one has 'points at infinity', these lie on the 'vanishing line'. The vanishing line consists of the points where pairs of parallel lines on the (flat) surface of the earth are depicted to meet. The need to introduce and study points at infinity led to projective geometry.

We start by considering the simplest kind of projections. If $\ell$ and m are two lines in the Euclidean plane $\mathbf{R}^2$, consider the problem of projecting one of them onto the other from a point $p \notin \ell \cup m$. This is done as follows:

If $x \in \ell$ and the line px meets m at y. The projection from $\ell$ to m is obtained by sending x to y.

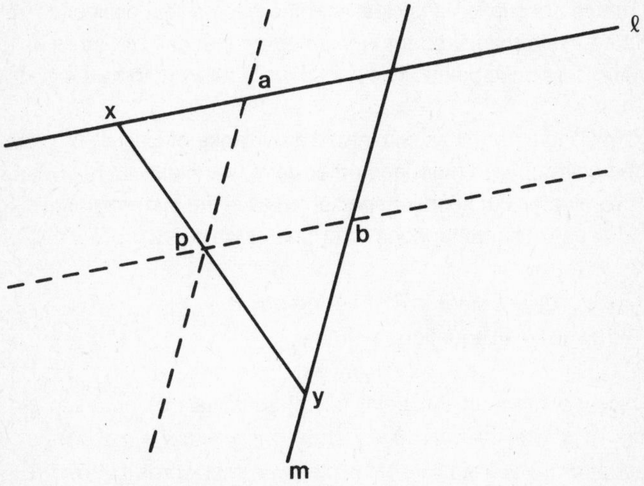


Unless $\ell$ and m happen to be parallel this map is not everywhere defined. If $x \in \ell$ is such that px is parallel to m ($x = a$ in the diagram), then there is no y. Similarly the point b is not in the image of the projection. As x tends to a, y tends to infinity and as x tends to infinity, y tends to b. If a point at infinity is added to each of $\ell$ and m then projection becomes well defined and bijective on the extended lines. These extended lines are called projective lines.

Similarly if $\pi_1$, $\pi_2$ are two planes in $\mathbf{R}^3$ and one wants to consider the projection of the plane $\pi_1$ from a point p to the plane $\pi_2$. Lines at infinity have to be introduced in order to make the projection well defined. The Euclidean plane together with the (projective) line at infinity forms the projective plane. The mathematician's 'line at infinity' corresponds to the artist's 'vanishing line'. This vanishing line is the image of the line at infinity of the plane of the earth under projection onto the artist's canvas.

**Homogeneous Co-ordinates**

The points of the projective line are those of $\mathbf{R}^1$ together with the point at infinity. For some purposes this description is adequate. (Compare with the treatment of the Riemann sphere in complex analysis. As we will see later, the Riemann sphere does fit in well with projective geometry.) A major drawback of this direct approach is that the point at infinity seems very different from the finite points. From another viewpoint this is not so. The points of $P^1$ all have the same standing, in other words $P^1$ is homogeneous. The co-ordinates that show this are the **homogeneous co-ordinates**. The points of $P^1$ are represented by ratios x/y of real numbers. If $y \neq 0$ one has $x/y \in \mathbf{R}^1$, and if $y = 0$, x/y is the point at infinity. More formally, a point in $P^1$ is an equivalence class of non-zero pairs (x,y) of real numbers under the equivalence relation: $(x_1,y_1) \sim (x_2,y_2)$ if and only if there is a non-zero $\lambda \in \mathbf{R}$ with $(x_1,y_1) = \lambda(x_2,y_2)$. So $P^1 = \mathbf{R}^2 \setminus \{0\}/\underline{x} \sim \lambda\underline{x}$. Usually the equivalence class containing (x,y) is denoted by [x:y]. This notation is meant to suggest that one is considering the ratio of x and y.

Similarly there are homogeneous co-ordinates for the projective plane $P^2$. A point is [x:y:z] and one has $[x:y:z] = [\lambda x:\lambda y:\lambda z]$ if $\lambda \neq 0$. As before $P^2 = \mathbf{R}^3 \setminus \{0\}/\underline{x} \sim \lambda\underline{x}$. The points with $z \neq 0$ have unique representatives of the form [x:y:1] (divide by z); such points correspond to the points $(x,y) \in \mathbf{R}^2$. The points with $z = 0$ are [x:y:0] and correspond to the points [x:y] of the (projective) line at infinity. From this description it is clear that there is no real distinction between the points at infinity in $P^2$ and the finite points, so that $P^2$ is homogeneous. The three points [1:0:0], [0:1:0] and [0:0:1] are regarded as base points. They are usually referred to as the vertices of the triangle of reference.

With this step from $\mathbf{R}^2$ to $P^2$ one has eliminated a nuisance of Euclidean geometry. This arises because in $\mathbf{R}^2$ some pairs of lines meet and other pairs (parallels) do not meet. In $P^2$ all pairs of lines meet and this eliminates the need for special cases in the statements of some theorems and proofs. To illustrate the use of homogeneous co-ordinates we give a proof of this fact.

**Proposition** Any pair of distinct lines in $P^2$ meet in a point.

**Proof** A line in $P^2$ is given by an equation

$$\ell x + my + nz = 0.$$

Note that if one representative of the point [x:y:z] satisfies this equation so does any other, $[\lambda x:\lambda y:\lambda z]$. Two lines $\ell_1 x + m_1 y + n_1 z = 0$ and $\ell_2 x + m_2 y + n_2 z = 0$ are distinct if one equation is not a multiple of the other, in other words if the matrix

$$A = \begin{bmatrix} \ell_1 & m_1 & n_1 \\ \ell_2 & m_2 & n_2 \end{bmatrix}$$

has rank 2. A point [x:y:z] lies on both lines if the vector

$$\underline{x} = \begin{bmatrix} x \\ y \\ z \end{bmatrix}$$

is a solution of $A\underline{x} = 0$. But as A has rank 2, the system $A\underline{x} = 0$ has a one dimensional solution, and all the non—zero solutions represent the same point in $P^2$.

This proof illustrates why some aspects of projective geometry are easier to study than the corresponding aspects of Euclidean geometry. One deals with homogeneous rather than inhomogeneous equations, and this makes the study of linear equations much simpler.

**Exercise** Show that an equation $f(\underline{x}) = 0$, where $\underline{x} = (x,y,z)$, defines a subset of $P^2$ if and only if $f(\underline{x}) = 0$ implies $f(\lambda\underline{x}) = 0$. Deduce that f must be homogeneous. Can the subset defined by f be empty?

**The Topology of $P^1$ and $P^2$**

We will first consider the projective line $P^1 = \mathbf{R}^1 \cup \{\infty\}$. The real line is homeomorphic to any open interval (for example, the map $x \to x/(1-|x|)$ defines a homeomorphism between $(-1,1)$ and $\mathbf{R}$). So $P^1$ is homeomorphic to an open interval with a point at infinity, in other words $P^1$ is homeomorphic to a closed interval with its two end points identified. This shows that $P^1$ is homeomorphic to the circle $S^1 = \{(x,y) \mid x^2+y^2=1\}$. By using stereographic projection one can, in a more direct fashion, set up a homeomorphism between $S^1$ and $\mathbf{R}^1 \cup \{\infty\}$. If $n \in S^1$ denotes the north pole – the point $(0,1)$, any line through n meets $S^1$ again in p and meets $\mathbf{R}$ in q. The map $p \to q$: $S^1 \setminus \{n\} \to \mathbf{R}$ is a homeomorphism and one can also send n to $\infty$. This defines a homeomorphism $S^1 \to \mathbf{R}^1 \cup \{\infty\} = P^1$.

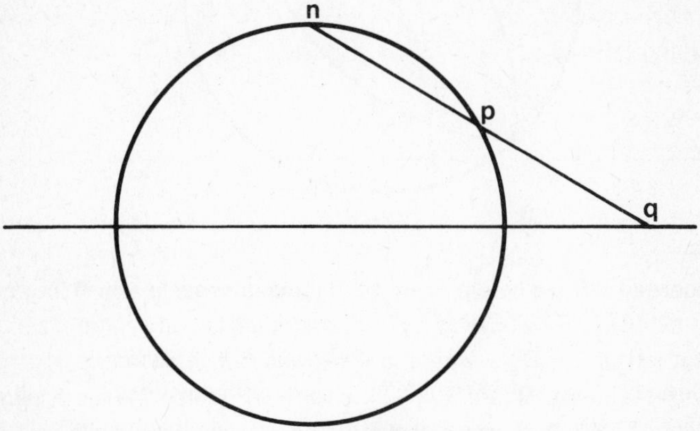


More formally, a homeomorphism can be set up by using homogeneous co-ordinates, every point of $P^1$ has a representative $[x{:}y]$ such that $x^2 + y^2 = 1$ (divide x and y by $(x^2+y^2)^{1/2}$). In fact, every point of $P^1$ has exactly two representatives of this form, namely $[x{:}y]$ and $[-x{:}-y]$. So $P^1$ is the space $S^1/\underline{x}\sim-\underline{x}$ and this defines its topology. There are two simple ways of seeing that the space $S^1/\underline{x}\sim-\underline{x}$ is itself homeomophic to $S^1$.

a) Every point of $S^1/\underline{x}\sim-\underline{x}$ has a representative in the lower semicircle, unique except for the two end points. When these two points are identified one gets a circle.

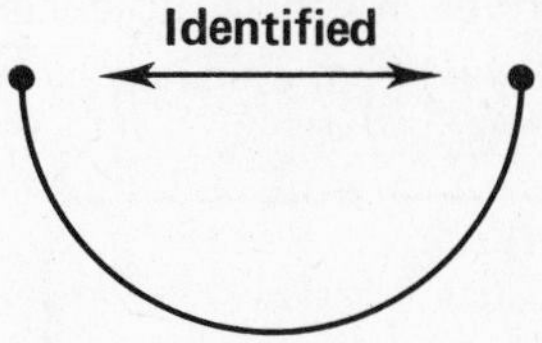


b) The map $z \to z^2$: $S^1 \to S^1$ (z a complex number with $|z| = 1$) identifies precisely the points z and $-z$. So the second circle can be regarded as the first circle modulo the identification $x \sim -x$.

To 'see' the line at infinity in

$$P^2 = \mathbf{R}^2 \cup \{\text{line at infinity}\}$$

choose a homeomorphism between $\mathbf{R}^2$ and the open disc $\mathring{D}^2 = \{\underline{x} \in \mathbf{R}^2 \mid \|\underline{x}\| < 1\}$; such a homeomorphism is given, in polar co-ordinates, by

$$(r,\theta)\rightarrow(r/(1-|r|),\theta)\colon \mathring{D}^2\rightarrow \mathbf{R}^2.$$

Two parallel lines $\ell$ and m in $\mathbf{R}^2$ when mapped into $D^2$ are as shown in the diagram.

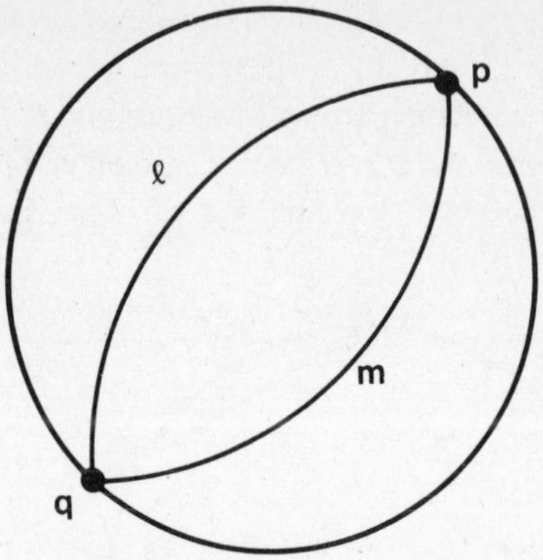


They 'want' to meet in p and q where $p = -q$, so to make a space in which they meet in only one point, one takes the disc $D^2 = \{(x,y)\colon x^2+y^2\leqslant 1\}$ and identifies antipodal pairs of points on the boundary. The space is $P^2 = D^2/\sim$ where $\underline{x} \sim \underline{y}$ if and only if either $\underline{x} = \underline{y}$ or $\underline{x} = -\underline{y}$ and $\|\underline{x}\| = 1$. The new points added to $\mathbf{R}^2$ are $S^1/\underline{x}\sim-\underline{x}$ which is a $P^1$, this is the line at infinity in $P^2$.

What we have done here is very reminiscent of our discussion of the topology of SO(3). In exactly the same way as was done there we see that $P^2$ is homeomorphic to $S^2/\underline{x} \sim -\underline{x}$. The lines in $P^2$ are the images of the great circles on the sphere $S^2 = \{(x,y,z) \in \mathbf{R}^3 | x^2 + y^2 + z^2 = 1\}$. The great circles being the intersections of $S^2$ with planes in $\mathbf{R}^3$ containing 0. We make all this formal by setting up the homeomorphism in homogeneous co-ordinates. From its description

$$P^2 = \mathbf{R}^3 \setminus \{0\}/\underline{x}\sim-\underline{x},$$

each point of $P^2$ is represented by precisely two (antipodal) points in $S^2$. So that $P^2 = S^2/\underline{x} \sim -\underline{x}$. One can define $P^n$ for all $n \geqslant 1$ as $S^n/\underline{x} \sim -\underline{x}$, and then the conclusion of our previous discussion about SO(3) is that SO(3) is homeomorphic to $P^3$.

The above description of $P^2$ as $D^2$ with antipodal points on the boundary identified coincides almost exactly with the description given in popular accounts of elementary topology. In such accounts the projective plane is obtained from a square by identifying edges according to the following diagam:

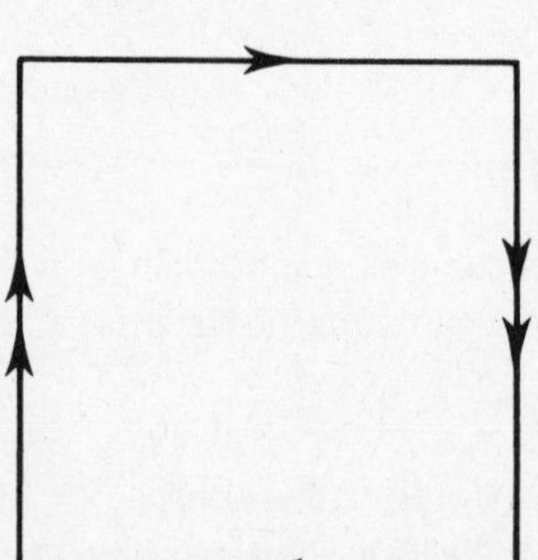

Under a suitable homeomorphism of the square with the disc, these identifications correspond exactly to the identification of antipodal points.

**Proposition** The space $P^2 \setminus \{x\}$ is homeomorphic to an open Möbius band for any $x \in P^2$. The space $P^2 \setminus \mathring{D}(x)$ is homeomorphic to a closed Möbius band for any open disc neighbourhood $\mathring{D}(x)$ of x.

**Proof 1** This is an elementary 'cutting and pasting' argument. One is allowed to cut the space as long as it is eventually pasted back together in the same way as it was cut.

It is clear that the two sentences of the proposition are essentially equivalent statements. For this first proof it is more convenient to prove the second sentence.

Consider $P^2 \setminus \mathring{D}(x)$, diagramatically we have:

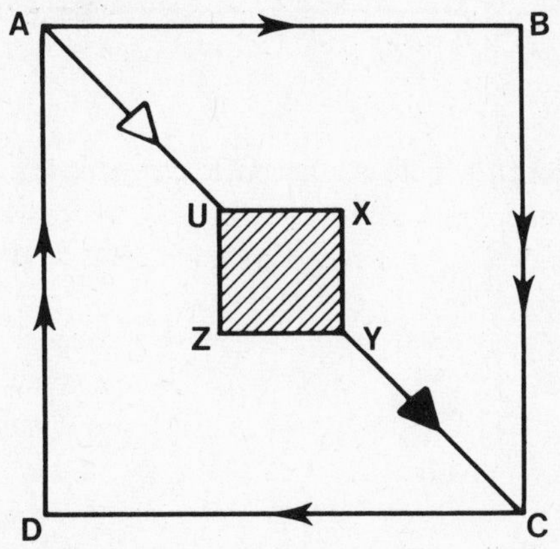


where the shaded region has been removed. If one cuts along AU and YC and sticks AD to BC there results the following.

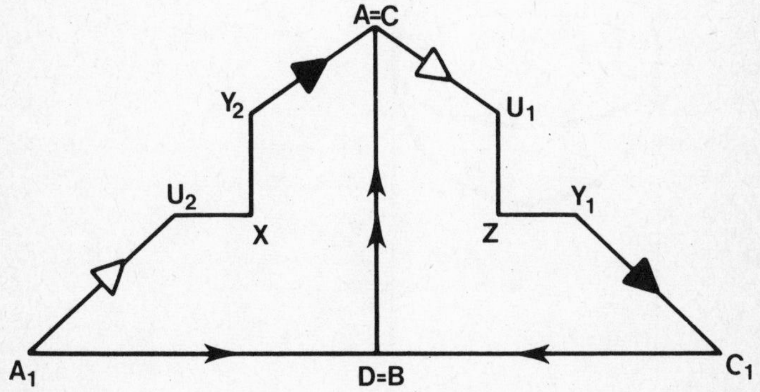

After sticking $A_1D$ to $C_1D$ this is clearly homeomorphic to

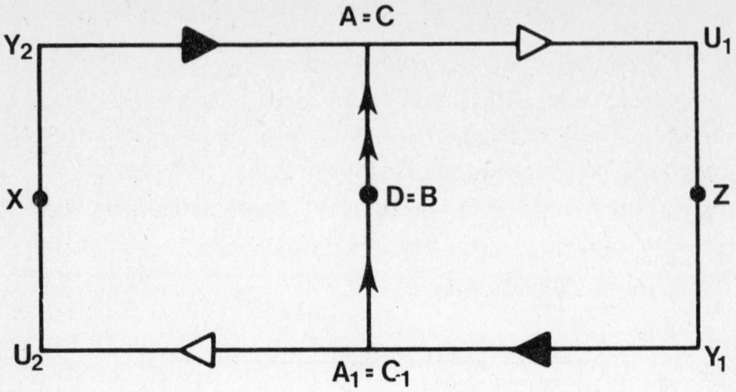


This is the Möbius band.

**Proof 2** For this second proof it is more convenient to prove the first sentence of the proposition. Consider $P^2 \setminus \{x\}$, this is homeomorphic to $S^2 \setminus \{n,s\}/\underline{a} \sim -\underline{a}$ where n and s are the north and south poles of the sphere $S^2$. But $S^2 \setminus \{n,s\}$ is homeomorphic to $S^1 \times \mathbf{R}^1$ and $\sim$ corresponds to the equivalence relation $(x,y) \sim (-x,-y)$.

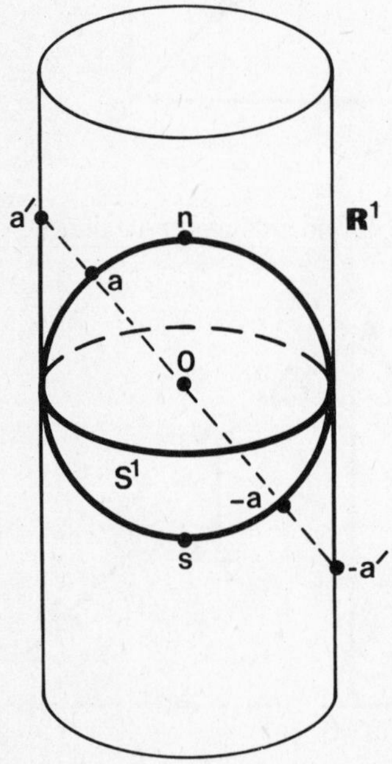


$$S^2 \setminus \{n,s\} \to S^1 \times \mathbf{R}^1$$
$$a \quad \to \quad a'$$

By the lemma on page 18, the space $S^1 \times \mathbf{R}^1/\sim$ is homeomorphic to the open Möbius band.

**Duality**

With no extra difficulty, one can define the projective space corresponding to a vector space V over any field F. It is

$$\mathbf{P}(V) = V \setminus \{0\}/\sim$$

where $u \sim v$ if and only if there is a (non-zero) $\lambda \in F$ with $u = \lambda v$. In our previous notation $P^1 = \mathbf{P}(\mathbf{R}^2)$ and $P^2 = \mathbf{P}(\mathbf{R}^3)$. Projective spaces of this type are much studied, but we will concentrate almost exclusively on lines and planes over the real and complex fields.

**Exercise** Define an embedding $V \subset \mathbf{P}(V \oplus F)$. What is the relationship between the projective subspaces of $\mathbf{P}(V \oplus F)$ and the affine subspaces of V?

With this elegant notation we can now explain one of the more beautiful aspects of projective geometry – duality.

If W is a subspace of V then one has an inclusion of the corresponding projective spaces $\mathbf{P}(V) \subset \mathbf{P}(W)$. A vector space V has a dual space $V^* = \mathrm{Hom}(V,F)$. This is the space of linear transformations from V to F (also called linear functionals on V). The projective spaces $\mathbf{P}(V)$ and $\mathbf{P}(V^*)$ are said to be dual. A point $p \in \mathbf{P}(V^*)$ determines a codimension one subspace W of V (that is W has dimension one less than V): Suppose p is represented by $\tilde{p} \in V^*$, then $W = \mathrm{Ker}\,\tilde{p}$, and as $\tilde{p}$ is non-zero, it follows that $\dim W = \dim V - 1$. It is easily verified that W does not depend on which representative $\tilde{p}$ of p was chosen. Hence a point of $\mathbf{P}(V^*)$ determines a codimension one projective subspace of $\mathbf{P}(V)$ and also vice versa, at least if V is finite dimensional, because then $V = V^{**}$. For example a point in $\mathbf{P}(\mathbf{R}^{3*})$ corresponds to a (projective) line in $\mathbf{P}(\mathbf{R}^3)$, and a line in $\mathbf{P}(\mathbf{R}^{3*})$ corresponds to a point in $\mathbf{P}(\mathbf{R}^3)$.

**Exercise** i) Show that three points in $\mathbf{P}(\mathbf{R}^{3*})$ are collinear if and only if the corresponding three lines in $\mathbf{P}(\mathbf{R}^3)$ are concurrent.

ii) Make corresponding statements in higher dimensions and for other fields.

An inner product x.y on $\mathbf{R}^n$ induces a linear isomorphism $\alpha: \mathbf{R}^n \to \mathbf{R}^{n*}$ by $\alpha(u)v = v.u$ and so $\mathbf{P}(\mathbf{R}^n)$ is then naturally isomorphic to $\mathbf{P}(\mathbf{R}^{n*})$. An easy consequence is that the space of codimension one subspaces in $\mathbf{P}(\mathbf{R}^n)$ is homeomorphic to $\mathbf{P}(\mathbf{R}^n)$ itself. In particular, the space of lines in $P^2$ is homeomorphic to $P^2$ itself. The lines in $P^2$ correspond to the lines in $\mathbf{R}^2$ together with the line at infinity, so that the space of lines in $\mathbf{R}^2$ is homeomorphic to $P^2 \setminus \{pt\}$. We have already come across this fact, but the present proof is more conceptual.

One of the main reasons that duality is useful is that every theorem has a dual. The dual of a true theorem is also true, so that only one of the two needs to be proved. As an example consider the theorem of Desargues:

**Theorem 16** If $p_1p_2p_3$ and $q_1q_2q_3$ are two triangles in $P^2$ with vertices $p_i$, $q_i$ such that the three lines $p_iq_i$ are concurrent then the three points $r_{ij} = p_ip_j \cap q_iq_j$ $(i \neq j)$ are collinear.

The dual is

**Theorem 16d** If $\ell_1\ell_2\ell_3$ and $m_1m_2m_3$ are two triangles in $P^2$ with sides $\ell_i$, $m_i$ such that the three points $\ell_i \cap m_i$ are collinear then the three lines $(\ell_i \cap \ell_j).(m_i \cap m_j)$ are concurrent.

In this case it happens that the dual theorem is also the converse of the original. Desargues theorem can be proved just using properties of $P^2$ but the simplest proof uses $P^3$.

**Proof** Assume $P^2 \subset P^3$, choose a projection point $O \notin P^2$ and suppose that the three lines $p_iq_i$ meet at v.

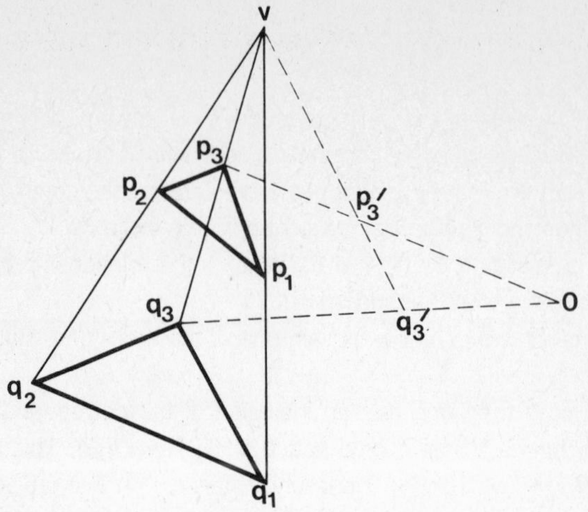


Choose some point $q_3' \neq q_3$ on $Oq_3$, the line $vq_3'$ meets $Op_3$ in $p_3'$ (as $vq_3'$ and $Op_3$ are coplanar). The triangles $p_1p_2p_3'$ and $q_1q_2q_3'$ lie in different planes in $P^3$, these planes meet in a line $\ell$.

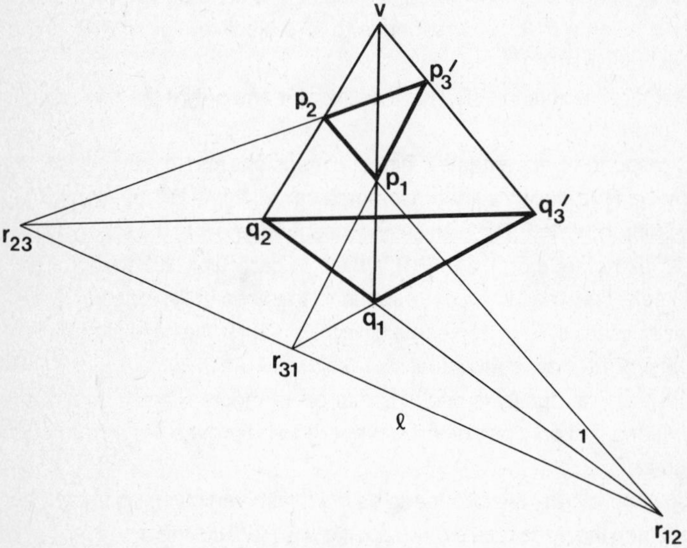


The points $r_{ij}$ are the projections of the points

$$p_1p_2 \cap q_1q_2, \qquad p_1p_3' \cap q_1q_3' \qquad \text{and} \qquad p_2p_3' \cap q_2q_3'$$

to $P^2$, hence the three points $r_{ij}$ lie on the projection of $\ell$ from O to $P^2$. This shows that the points $r_{ij}$ are collinear.

By applying the theorem to the dual $P^2$ one has simultaneously proved the dual and hence (in this case) the converse.

**Exercise** By choosing a suitable line at infinity deduce the following: If $p_1p_2p_3$, $q_1q_2q_3$ are two triangles in $\mathbf{R}^2$ with the three lines $p_iq_i$ parallel, show that either the three points $p_ip_j \cap q_iq_j$ are collinear, or the three pairs of lines $p_ip_j$, $q_iq_j$ are parallel.

**Projective Groups**

In the introduction to these notes the importance of a group acting on a geometry was stressed, and the reader should now be familiar enough with projective geometry for the group to be introduced.

If $T: V \to V$ is a linear isomorphism, it clearly induces an isomorphism $\mathbf{P}(V) \to \mathbf{P}(V)$. Some linear isomorphisms of V induce the identity on $\mathbf{P}(V)$. If $T(v) = \lambda v$ for a scalar $\lambda$ then one certainly gets the identity on $\mathbf{P}(V)$, because one has divided out by the scalars in forming $\mathbf{P}(V)$. Scalar multiplications are the only linear isomorphisms of V that induce the identity on $\mathbf{P}(V)$. When $V = F^n$, its group of linear isomorphisms is $GL(n,F)$.

The quotient group $PGL(n,F) = GL(n,F)/\{\lambda I\}$ is called the **projective group** and its elements are called **projectivities**.

We will illustrate this definition by considering the case $n = 2$ and $F = \mathbf{C}$ in some detail. You might have come across this already in complex analysis. To conform with a standard notation we will write $P^{n-1}(F)$ for $\mathbf{P}(F^n)$.

**Example** $P^1(\mathbf{C})$ is the Riemann sphere because $P^1(\mathbf{C})$ is $\mathbf{C}$ together with a point at infinity. This latter description is the usual description of the Riemann sphere. The points of $P^1(\mathbf{C})$ are, in homogemeous co-ordinates, pairs $[z:w]$ where $(z,w) \neq (0,0)$. If $w \neq 0$, $[z:w]$ corresponds to the point $\zeta = z/w \in \mathbf{C}$, and if $w = 0$, it is the point at infinity,

Stereographic projection shows that $P^1(\mathbf{C})$ is homeomorphic to the sphere $S^2 = \{\underline{x} \in \mathbf{R}^3 | \, \|\underline{x}\| = 1\}$.

Stereographic projection is the map $p \to q$ where npq are collinear. It can also be described as the map obtained by rotating the map $S^1 \to P^1(\mathbf{R})$, defined on page 53, about the vertical axis. Its properties are discussed in more detail in Part III.

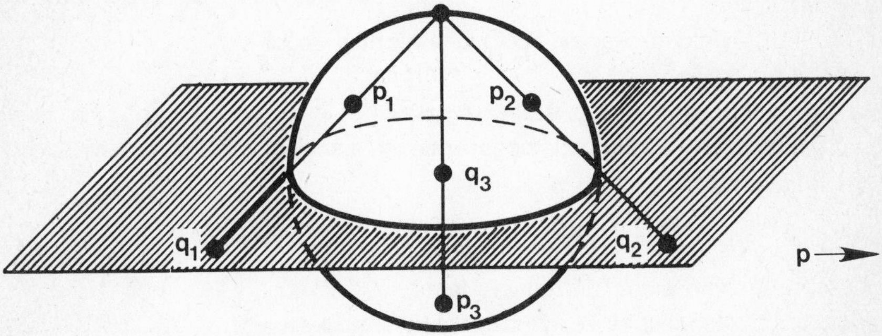


To understand the action of $PGL(2,\mathbf{C})$ on $P^1(\mathbf{C})$ one must take representatives,

$$\begin{bmatrix} a & b \\ c & d \end{bmatrix} \in GL(2,\mathbf{C}) \text{ and } \begin{bmatrix} z \\ w \end{bmatrix} \in \mathbf{C}^2 \setminus \{0\},$$

of the elements of $PGL(2,\mathbf{C})$ and $P^1(\mathbf{C})$ under consideration. One gets

$$\begin{bmatrix} a & b \\ c & d \end{bmatrix} \begin{bmatrix} z \\ w \end{bmatrix} = \begin{bmatrix} az+bw \\ cz+dw \end{bmatrix}$$

that is, the point $[az+bw:cz+dw] \in P^1(\mathbf{C})$. If $w = 0$, let $\zeta = z/w$, then this transformation is the **fractional linear (or Möbius) transformation**

$$\zeta \to \begin{bmatrix} a\zeta+b \\ c\zeta+d \end{bmatrix} \quad \text{for } \zeta \neq -d/c.$$

The point at infinity is mapped to a/c and the point $-d/c$ is mapped to the point at infinity. (Remember that we are assuming $ad - bc \neq 0$.) This transformation is clearly a self homeomorphism of $P^1(\mathbf{C})$.

Unless $c \neq 0$ it is not a well defined map from **C** to itself. Projective space here plays the same role as it did in our discussion of projections (on page 51) – it allows certain interesting maps to be well defined.

When handling projective lines, it is often convenient not to have two variables z, w but merely to have one variable $\zeta$ whose values are in $\mathbf{C} \cup \{\infty\}$. One uses some 'rules of thumb' when handling the case $\zeta = \infty$; these can be proved using the more rigorous approach using the two variables z,w: for example $(a\infty+b)/(c\infty+d)$ is a/c.

If f: $P^1(\mathbf{C}) \to P^1(\mathbf{C})$, it makes sense to say that f is analytic at any point of $P^1(\mathbf{C})$. In particular, one can consider analytic maps f: $P^1(\mathbf{C}) \to P^1(\mathbf{C})$ which are bijections and whose inverses are also analytic, such maps are called **bianalytic isomorphisms**. A Möbius transformation is a bianalytic isomorphism and there are no others:

**Theorem 17** The group of bianalytic isomorphims f: $P^1(\mathbf{C}) \to P^1(\mathbf{C})$ is PGL(2,**C**).

**Proof** Suppose that f is a bianalytic isomorphism. Then there is a single point $a \in P^1\mathbf{C}$ such that $f(a) = \infty$. By composing with a Möbius transformation, we can assume $a = 0$. Let f have residue $\alpha$ at 0, then

$$f(z) = g(z) + \alpha/z,$$

where g(z) is analytic and finite valued everywhere on $P^1\mathbf{C}$. By Liouville's theorem, g is a constant c. Hence

$$f(x) = c + \alpha/z$$

and so f is a Möbius transformation.

Two important properties of Möbius transformations are:

**Theorem 18** If f is a Möbius transformation, then f is conformal, that is, it preserves angles, and f sends circles to circles.

**Proof** The map f is conformal because any bianalytic map is conformal.

The equation of a circle is

$$Az\bar{z} + Bz + \bar{B}\bar{z} + C = 0$$

where A, C are real. If $A = 0$ the circle degenerates to a straight line, that is, a circle through $\infty$. Let $z = (aw+b)/(cw+d)$ then w satisfies

$$A\frac{(aw+b)(\bar{a}\bar{w}+\bar{b})}{(cw+d)(\bar{c}\bar{w}+\bar{d})} + B\frac{(aw+b)}{(cw+d)} + \bar{B}\frac{(\bar{a}\bar{w}+\bar{b})}{(\bar{c}\bar{w}+\bar{d})} + C = 0$$

which reduces to an expression of the required form.

Projections are projectivities as we will now show for the case $n = 1$.

**Example** A projection f: $P^1(F) \to P^1(F)$ is an element of PGL(2,F).

We consider two lines $\ell_1, \ell_2$ in $P^2(F)$ and a point p not in either. By choosing a suitable line at infinity, we may assume that $\ell_1, \ell_2$ meet in $F^2$ and that $p \in F^2$. Further by choosing a suitable basis for $F^2$ one can assume that $\ell_1$ is $x = 0$, $\ell_2$ is $y = 0$ and that p is the point (1,1). The map $\ell_1 \to \ell_2$ is then defined by $f(x) = y$ if (x,0), (0,y) and (1,1) are collinear; that is, if $y = (x-1)/x$. So $f(x) = (x-1)/x$ and f is in PGL(2,F).

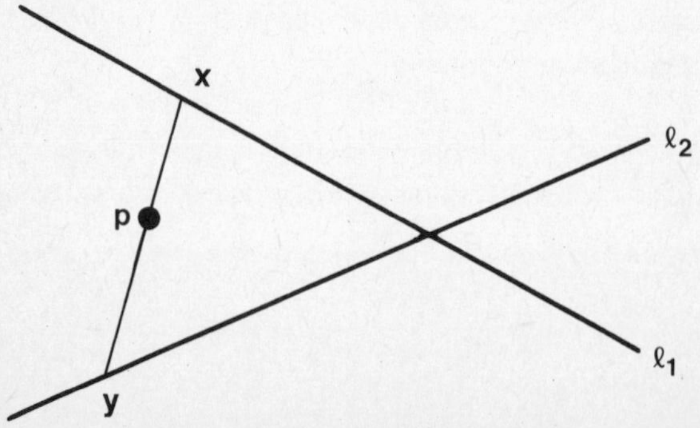

We now study how the group PGL(2,F) acts on $P^1(F)$:

**Theorem 19** Given two sets $\{p,q,r\}$; $\{p^1,q^1,r^1\}$ each consisting of three distinct points in $P^1(F)$. There is a unique $f \in PGL(2,F)$ such that $fp = p^1$, $fq = q^1$ and $fr = r^1$ (such an action is called triply transitive).

**Proof** We will use the following lemma to choose good representatives in $F^2$ for the points in $P^1(F)$.

**Lemma** Given three distinct points p, q, $r \in P^1(F)$, they have representatives $\tilde{p}, \tilde{q}, \tilde{r} \in F^2$ such that $\tilde{p} = \tilde{q} + \tilde{r}$.

**Proof of lemma** Choose any three representatives $\tilde{p}_1$, $\tilde{q}_1$, $\tilde{r}_1$. As q, r are distinct, $\tilde{q}_1$ and $\tilde{r}_1$ are independent and therefore span $F^2$. Hence there are constants $\lambda$, $\mu$ such that $\tilde{p}_1 = \lambda\tilde{q} + \mu\tilde{r}_1$. The representatives we need are then $\tilde{p} = \tilde{p}_1$, $q = \lambda\tilde{q}_1$ and $\tilde{r} = \mu\tilde{r}_1$.

Using this lemma choose representatives $\tilde{p}$, $\tilde{q}$, $\tilde{r}$ and $\tilde{p}^1$, $\tilde{q}^1$, $\tilde{r}^1$ such that $\tilde{p} = \tilde{q} + \tilde{r}$ and $\tilde{p}^1 = \tilde{q}^1 + \tilde{r}^1$. There is a non-singular linear transformation $\tilde{f}$: $F^2 \to F^2$ defined by $\widetilde{fq} = \tilde{q}^1$ and $\widetilde{fr} = \tilde{r}^1$ (remember that $\{\tilde{q},\tilde{r}\}$ and $\{\tilde{q}^1,\tilde{r}^1\}$ are bases for $F^2$). Clearly $\widetilde{fp} = \tilde{p}^1$, and so the image of $\tilde{f}$ in PGL(2,F) has the required property.

To prove uniqueness, it is easy to see that one needs only to verify that, if p, q, $r \in P^1(F)$ are distinct points such that $fp = p$, $fq = q$ and $fr = r$, then f is the identity. Choose representatives $\tilde{f} \in GL(2,F)$, $\tilde{p}$, $\tilde{q}$, $\tilde{r} \in F^2$ such that $\tilde{p} = \tilde{q} + \tilde{r}$. As $fp = p$ etc., there are $\lambda$, $\mu$, $\nu \in F$ such that $\widetilde{fp} = \lambda\tilde{p}$, $\widetilde{fq} = \mu\tilde{q}$ and $\widetilde{fr} = \nu\tilde{r}$. So

$$\lambda(\tilde{q}+\tilde{r}) = \lambda\tilde{p} = \widetilde{fp} = \tilde{f}(\tilde{q}+\tilde{r}) = \widetilde{fq} + \widetilde{fr} = \mu\tilde{q} + \nu\tilde{r}.$$

But $\tilde{q}, \tilde{r}$ are independent so $\lambda = \mu = \nu$. So $\tilde{f} = \lambda I$ and hence f is the identity in PGL(2,F).

This theorem can be used to introduce an important invariant of projective geometry.

## The Cross-Ratio

The three points 0, 1, $\infty \in P^1(F)$ are called the standard reference points. In homogeneous co-ordinates they are [0:1], [1:1] and [1:0].

Given any four distinct points p, q, r, $s \in P^1(F)$, Theorem 19 provides us with a unique $f \in PGL(2,F)$ such that $fp = \infty$, $fq = 0$, $fr = 1$, the image of the fourth point s is then determined, say $\lambda = fs$, and is called the **cross-ratio** of p,q,r,s. It is usually written as $\lambda = (pq,rs)$ or $(p,q;r,s)$. This cross-ratio can be defined directly in terms of the points p, q, r and s as follows: Define the arithmetic rules on $P^1(F)$ by using the usual ones on F and treating $\infty$ using the 'rule of thumb' mentioned on page 60; that is, write $\infty$ as 1/0 and then multiply throughout by 0. This could be formalised but it is easier to do it informally. With this convention, the cross-ratio can be evaluated numerically as

$$(pq,rs) = \frac{(p-r)(q-s)}{(p-s)(q-r)}$$

**Proposition** The two definitions of the cross-ratio agree.

**Proof** If p, q, r, $s \in P^1(F)$ and $f \in PGL(2,F)$ we need to show that $(fp,fq;fr,fs) = (p,q;r,s)$. Let $p = [p_0:p_1]$ etc., then an easy calculation shows that the cross-ratio $(p,q;r,s)$ is

$$\frac{\det(p,r)\det(q,s)}{\det(p,s)\det(q,r)} \text{ where } (p,r) \text{ is the matrix } \begin{bmatrix} p_0 & r_0 \\ p_1 & r_1 \end{bmatrix}$$

Note that det(p,r) is not well-defined but that the cross-ratio is well defined: If $\tilde{f} \in GL(2,F)$ represents f then fp is the point $[p_0{}^1:p_1{}^1]$ where $\begin{bmatrix} p_0{}^1 \\ p_1{}^1 \end{bmatrix} = \tilde{f}\begin{bmatrix} p_0 \\ p_1 \end{bmatrix}$ so $\det(fp,fr) = \det\tilde{f}.\det(p,r)$. The result now follows.

**Fixed Points of Projectivities**

Finding the fixed points of a map $f\colon P^n(F) \to P^n(F)$ for $f \in PGL(n+1,F)$ is the same as finding the eigenvectors of a representative $\tilde{f} \in GL(n+1,F)$ because $fp = p$ if and only if there is a $\lambda$ with $\tilde{f}\tilde{p} = \lambda\tilde{p}$ for a representative $\tilde{p}$ of p. Projectivities are sometimes named according to their fixed point sets. The reader may verify that the following cover all the possibilities in each of the stated cases:

<u>$n = 1, F = \mathbf{C}$.</u> There are two types of elements apart from the identity.

**Parabolic:** Exactly one fixed point e.g. $\begin{bmatrix} 1 & 1 \\ 0 & 1 \end{bmatrix}$ (not diagonalizable)

**Hyperbolic:** Exactly two fixed points e.g. $\begin{bmatrix} 1 & 0 \\ 0 & 2 \end{bmatrix}$ (diagonalizable)

<u>$n = 1, F = \mathbf{R}$</u> There are three types of elements apart from the identity; these are distinguished using $\triangle = \det \begin{bmatrix} a & b \\ c & d \end{bmatrix}$ and $t = \operatorname{trace} \begin{bmatrix} a & b \\ c & d \end{bmatrix}$

**Elliptic:** No fixed point $t^2 < 4\triangle$

**Parabolic:** Exactly one fixed point $t^2 = 4\triangle$

**Hyperbolic:** Exactly two fixed points $t^2 > 4\triangle$.

**The Elliptic Plane**

Thus far we have considered all the projective transformations of a projective space $P^n(F)$. When $F = \mathbf{R}$ one can define a good metric on $P^n(\mathbf{R}) = P^n$, and then consider the group of isometries. We will do this in the case $n = 2$. It may be helpful to have the following general remarks about lines in metric spaces:

A **curve** (or path) in a metric space X is the image of the unit interval [0,1] under a continuous map $[0,1] \to X$. A **geodesic** in X is a curve of shortest length between two points of X. This is a naive form of a more complicated definition, and is not quite rigorous but it is good enough to convey the general ideas. One has to define geodesics locally and then extend these globally. For example consider the sphere $S^2$, on a great circle both the long curve from a to b and the short curve are local geodesics, but according to our definition only the short curve is a geodesic. One can think of the geodesics as the straight lines in the space.

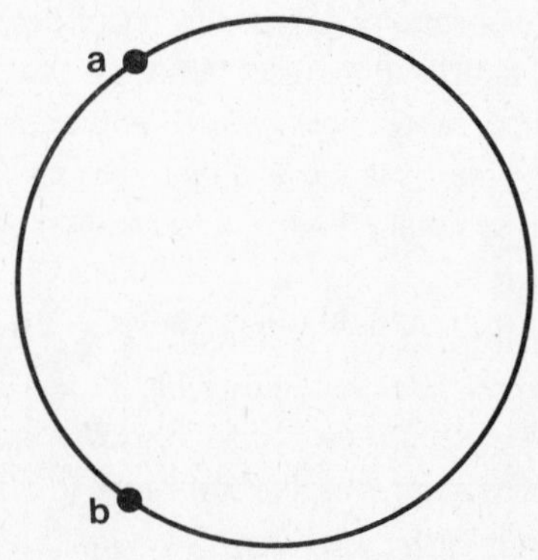

The geodesics on the sphere $S^2$ are the great circles – the intersections of linear planes in $\mathbf{R}^3$ with $S^2$ (as is well known to navigators!). One can therefore consider a geometry on $S^2$ on which the lines are these great circles. The main defect with this as a geometry is that any two lines meet in two (antipodal) points. However the geometry is quite tractable, calculations can be performed by using 'spherical trigonometry'. One can get rid of the defect by identifying all pairs of antipodal points with each other, $x \sim -x$. Thus one obtains the projective plane with a metric – this is called the **elliptic plane** and the geodesics are precisely the projective lines. As the sphere $S^2$ has unit radius, one can easily calculate that the total area of the elliptic plane is $2\pi$ (half the area of $S^2$, because of the identifications) and that the length of a projective line is $\pi$.

The group of isometries of the elliptic plane consists of those elements of PGL(3,**R**) that preserve the metric. Any isometry of $P^2$ arises from an isometry of $S^2$ and the isometries of $S^2$ make up the group O(3). Hence the group of isometries of the elliptic plane is the image of $O(3) \subset GL(3,\mathbf{R})$ in PGL(3,**R**); call this PO(3). The only scalar matrices in O(3) are $\pm I$ so $PO(3) \cong O(3)/\{\pm I\}$. As $-I \notin SO(3)$ and $O(3) \cong SO(3) \times \{\pm I\}$, we see that PO(3) is isomorphic to SO(3). Note however that the groups PO(n) and SO(n) are not always isomorphic; PO(n) has two components for even n. In particular, the isometries of $P^1$ form the group $O(2)/\{\pm I\}$. The elements in the component $SO(2)/\{\pm I\}$ correspond to rotations of $S^1$ and have no fixed point (except that the identity has fixed points), so this component consists entirely of elliptic elements. The other component contains reflections of $S^1$ and each element has exactly two fixed points in $P^1$ and so is hyperbolic:

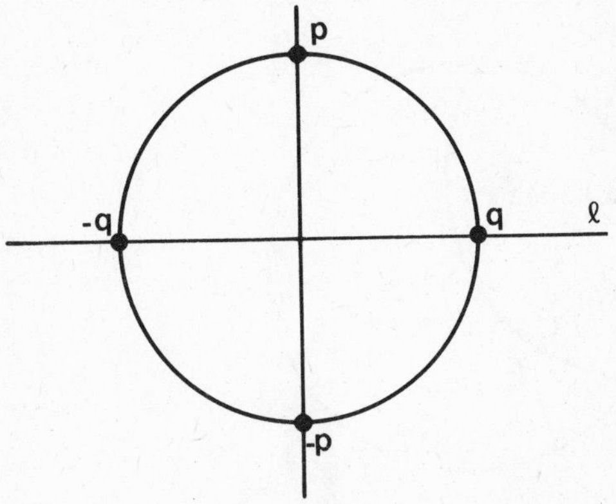


Reflection in ℓ

The two fixed points are the images of p and q. Any parabolic element of PGL(2,**R**) is not an isometry.

As an example of a calculation in the elliptic plane we have

**Example** Let $\triangle$ be a triangle in the elliptic plane whose angles are $\alpha$, $\beta$ and $\gamma$. Its area is $(\alpha+\beta+\gamma)-\pi$.

$\triangle$ is bounded by (segments of) three projective lines. Consider the corresponding three great circles ℓ, m, n in $S^2$.

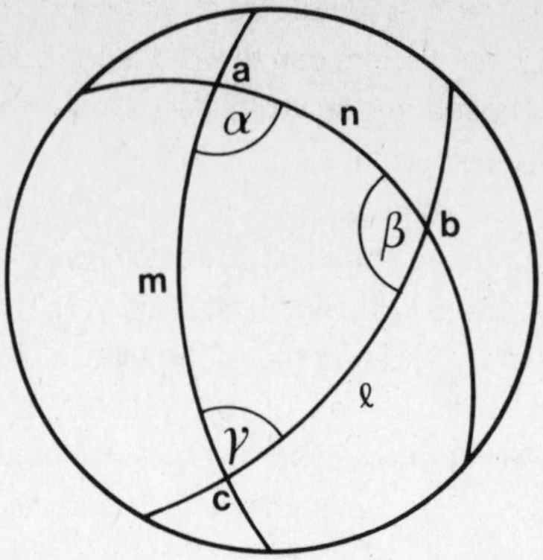


The area of the elliptic triangle $\triangle$ clearly equals the area of the spherical triangle abc. The area of the 'lune' between the lines m and n is $2\alpha$.

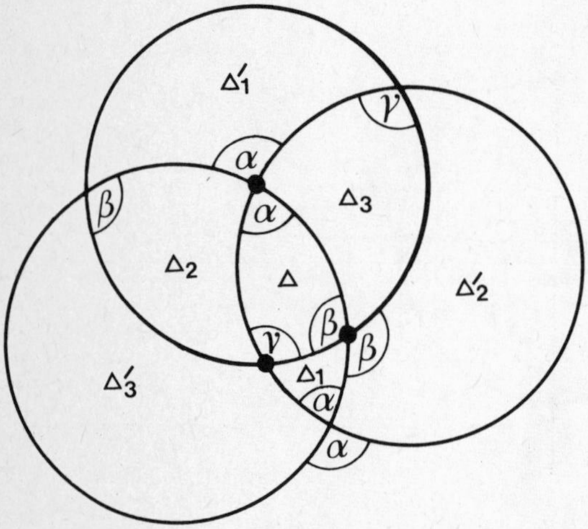

Calculating, one gets

$\triangle + \triangle_1 = 2\alpha$

$\triangle + \triangle_2 = 2\beta$

$\triangle + \triangle_3 = 2\gamma$

$\triangle + \triangle_1 + \triangle_2 + \triangle_3 = 2\pi$

So $2\triangle + 2\pi = 2\alpha + 2\beta + 2\gamma$

giving the result.

**Exercise** Calculate the area of a 'convex' spherical polygon in terms of its angles.

**Conics**

Conics have been studied since the time of the ancient Greeks. They seem to have been invented to solve the Delian problem: to find a geometrical method of constructing a number x such that $x^3 = 2$. It can be proved, using a little of the theory of fields, that such an x cannot be constructed by ruler and compass alone. Consider the two parabolae $x^2 = y$ and $y^2 = 2x$. The x-coordinate of their (non zero) intersection satisfies $x^3 = 2$. This is one of the solutions of the Delian problem given by Menaechmus in the 4th century BC. The Greeks gave numerous other solutions of this problem that used conics in various ways. For these and other interesting uses of conics see the book '100 Great Problems of Elementary Mathematics' by Dörrie (Dover publications).

Conics, as their full name conic sections suggests, are obtained by considering the planar sections of a cone. For example, if C is the cone $x^2 + y^2 = z^2$ in $\mathbf{R}^3$ whose vertex V is 0, one can consider $C \cap \pi$ for various planes $\pi$ in $\mathbf{R}^3$:

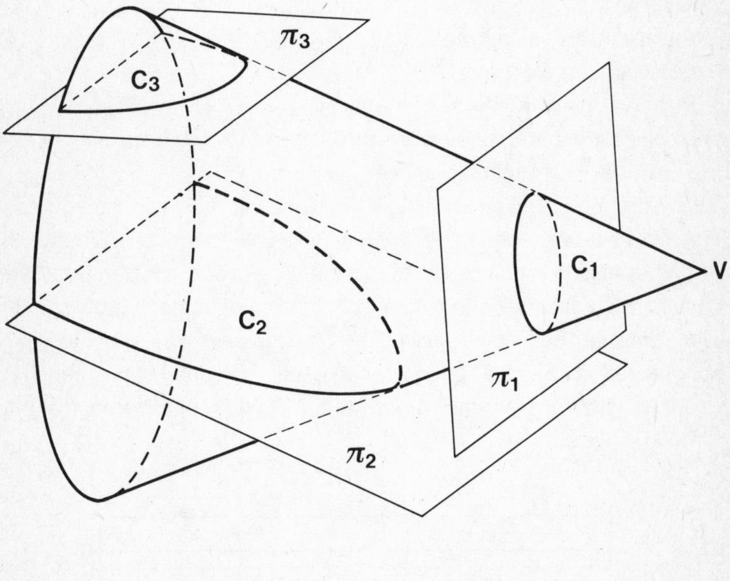

We list the various possibilities for $\pi$:

I. If $V \in \pi$, we get the singular conics:

a. $\pi$ meets C only at V, the conic is one point.

b. $\pi$ is tangent to C along a generator, the conic is a straight line.

c. $\pi$ meets C in two straight lines.

II. If $V \notin \pi$, there are three cases corresponding to which of the three possibilities of I above holds for the plane $\pi^1$ parallel to $\pi$ and passing through V. These are the non-singular conics.

a. $\pi$ meets C in an ellipse ($C_1$ in diagram).

b. $\pi$ meets C in a parabola ($C_2$ in diagram).

c $\pi$ meets C in a hyperbola ($C_3$ in diagram).

**Exercise** Let $D^3$ be a solid ball in $\mathbf{R}^3$, $p \in \mathbf{R}^3 \setminus D^3$ and $\pi$ a plane in $\mathbf{R}^3$ not containing p and not meeting $D^3$. What shape is the projection of $D^3$ from p onto the plane $\pi$?

From the projective viewpoint all the non-singular conics are the same, because they can clearly be projected to each other from the vertex V – see the above picture. It is therefore easier to study conics from the projective viewpoint. There is only one type, the difference being only in the choice of the line at infinity. There are three possible choices relative to the conic C:

i) ellipse, the line at infinity does not meet C

ii) parabola, the line at infinity touches C

iii) hyperbola, the line at infinity meets C in two points.

Theorems about conics are therefore best proved in $P^2$ and to obtain special theorems about one of the three types in $\mathbf{R}^2$ the line at infinity is chosen carefully in $P^2$.

By suitably choosing the line at infinity, a non-singular conic in $P^2$ can be regarded as an ellipse in $\mathbf{R}^2$. Hence any non-singular conic in $P^2$ is homeomorphic to a circle. We have seen that any projective line in $P^2$ is also homeomorphic to a circle. However a line can be distinguished from a conic topologically, because their complements in $P^2$ are not homeomorphic – they lie inside $P^2$ in different ways. This is not easy to visualise directly because $P^2$ itself is not easily visualised, but it is worth making the effort. A similar phenomenon occurs with the Möbius band M: consider the 'middle circle' $\ell$ of M which goes round M once and a circle m that is very close to the edge of M and so goes round M twice.

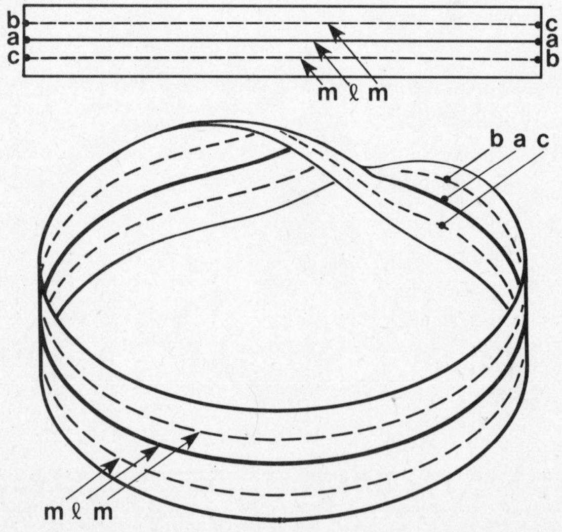

The complement of $\ell$ is connected and is homeomorphic to an annulus but the complement of m has two components. One is homeomorphic to an annulus and the other to a Möbius band. (If you find this hard to visualise you can cut two paper Möbius bands, one along $\ell$ and the other along m.) Remember the proposition on page 55 that $P^2 \setminus \{\text{open disc}\}$ is homeomorphic to the Möbius band. A line in $P^2$ is like the line $\ell$ of the Möbius band and a conic in $P^2$ is like the line m of the Möbius band. In fact, after removing a point not on the line or conic the situation is exactly like this, to see this we use a slightly different viewpoint:

The space $\mathbf{R}^2 \setminus \{\text{line}\}$ is disconnected but $P^2 \setminus P^1$ is connected. Consider the sphere $S^2$ with a great circle removed. Under the map $\pi: S^2 \to P^2$ this space is mapped to $P^2 \setminus P^1$ and consists of two hemispheres which are identified under $\pi$ so $P^2 \setminus P^1$ is homeomorphic to an open disc and therefore to $\mathbf{R}^2$. Alternatively, think of $P^1 \subset P^2$ as the line at infinity, removing it leaves $\mathbf{R}^2$. The space $P^2 \setminus \{\text{conic}\}$ is disconnected, one component (the 'inside') is homeomorphic to $\mathbf{R}^2$ the other component (the 'outside') is homeomorphic to an open Möbius band. To see this, choose the line at infinity to miss the conic so that the conic is an ellipse in $\mathbf{R}^2$, its inside is homeomorphic to $\mathbf{R}^2$ and its outside is homeomorphic to an annulus. But its outside in $P^2$ is homeomorphic to $P^2 \setminus \{\text{disc}\}$ which is a Möbius band.

The equation defining a conic in $\mathbf{R}^2$ is a polynomial of degree two

$$ax^2 + by^2 + 2dy + 2ex + 2fxy + c = 0.$$

An equation defining a subset of a projective space must be homogeneous otherwise some co-ordinates representing a point may satisfy the equation and others may not. To obtain a homogeneous equation from the above equation, one multiplies each term with various powers of z so that each term has degree two giving the equation

$$q(x,y,z) \equiv ax^2 + by^2 + cz^2 + 2dyz + 2ezx + 2fxy = 0$$

(More properly we should have taken new co-ordinates $\xi, \eta, \zeta$ with $x = \xi/\zeta$, $y = \eta/\zeta$ giving the co-ordinates $[\xi:\eta:\zeta]$ in $P^2$ and (x,y) in $\mathbf{R}^2$ with $\zeta = 0$ being the line at infinity.)

In matrix notation

$$q(x,y,z) \equiv \underline{x}^t A \underline{x}$$

where x is the vector $\begin{bmatrix} x \\ y \\ z \end{bmatrix}$ and A is the matrix $\begin{bmatrix} a & f & e \\ f & b & d \\ e & d & c \end{bmatrix}$

An easy check shows that the conic is non-singular if and only if the matrix A is non-singular. With this algebraic definition one can discuss conics in $\mathbf{P}^2(F)$ for any field F. They are non-singular if the matrix A is non-singular.

**Exercise** (For those with a knowledge of the calculus of several variables.) The projective plane $P^2$ is a manifold (indeed so is $P^n$ for any n). Show that the conic C is non-singular if and only if the function q(x,y,z) has non-vanishing derivative at each point of C and that then C is a manifold.

Algebraically a conic is given by a quadratic form in three variables, so that conics can be studied using the theory of quadratic forms.

## Diagonalization of Quadratic Forms

The algebraic invariants of quadratic forms arise from the attempt to diagonalise them. We will do this for the fields **R** and **C** but it can be done over other fields. The interested reader can do this himself.

**Over R** The invariants of a real quadratic form are its rank and signature. Given the rank and signature, there is a linear change of variables which transforms the quadratic form to a canonical form. In our case this linear change of co-ordinates gives a projective transformation of $P^2$ that sends q to a canonical form. This canonical form is easy to analyse and we summarize the results in the following table. As we are considering the points where q vanishes, the sign of q is immaterial so we need only consider quadratic forms q with positive signatures.

| Rank | Signature | Canonical Form | Point Set of Conic |
|---|---|---|---|
| 3 | 3 | $x^2+y^2+z^2=0$ | Empty |
| 3 | 1 | $x^2+y^2=z^2$ | Non-singular conic |
| 2 | 2 | $x^2+y^2=0$ | One point: [0:0:1] |
| 2 | 0 | $x^2=y^2$ | Two lines, $x=\pm y$ They meet at [0:0:1] |
| 1 | 1 | $x^2=0$ | One line, $x=0$, of multiplicity 2, "double line" [0:a:b] |

It is worth noting that the three singular cases remain distinct in $P^2$ and that the empty conic does not arise as a conic section unless the cone itself degenerates to a line.

**Over C** In this case the only invariant is the rank.

| Rank | Canonical Form | Point Set of Conic |
|---|---|---|
| 3 | $x^2+y^2+z^2=0$ | Non-singular conic |
| 2 | $x^2+y^2=0$ | Two lines, $x=\pm iy$ |
| 1 | $x^2=0$ | One line, $x=0$, of multiplicity 2 "double line" |

The classification becomes much simpler on passing from the Euclidean to the projective case and again on passing from the real to the complex case. A good way of understanding the classification is to go backwards through these stages. That is why complex projective geometry was so extensively studied in the last century –theorems in the other situations being proved by specialising theorems of $P^2(\mathbf{C})$. Of course the amount gained by doing this for conics is not great, but in the study of higher degree curves (the zero sets of polynomials of degree $\geqslant 3$) the advantages are much more substantial. Newton was the first to attempt to classify cubics (curves of degree 3) in $\mathbf{R}^2$ and he obtained 72 of the possible 78 forms. In $P^2(\mathbf{C})$ there are just 3 forms.

Now we investigate the space of all conics in $P^2(\mathbf{C})$ (this works over any field but there is no significant loss in just thinking about the case $F=\mathbf{C}$).

**Theorem 20** The space of all conics in $P^2(\mathbf{C})$ is $P^5(\mathbf{C})$.

**Proof** This is immediate because a conic is determined by the six numbers a, b, c, d, e, f and multiplying throughout by a non-zero scalar gives the same conic.

Note that we are really saying that the space of quadratic forms in three variables, up to scalar multiplication, is $P^5(F)$. These forms do not always correspond to different point sets in $P^2(F)$. For example, over **R** there are many different quadratic forms that give the empty point set. However for an 'algebraically closed' field such as **C** the correspondence of the theorem is also an exact one for point sets.

**Exercise** Let $v\colon P^2(F)\to P^5(F)$ be the Veronese map, given by

$$v[x:y:z]=[x^2:y^2:z^2:2yz:2zx:2xy]$$

then the images of the conics in $P^2(F)$ are the intersections of the image of v with the various hyperplanes of $P^5(F)$ (note that v is always injective). This gives a correspondence between the set of conics in $P^2(F)$ and the dual of the $P^5(F)$ in which the image of v lies. The image of the map v is called the Veronese surface.

The various types of conics will form certain subsets of $P^5(F)$, which we now investigate when $F = \mathbf{C}$.

1. The **rank one** conics correspond to the matrices of rank one and so are given by the subspace of $P^5(\mathbf{C})$ defined by the equations:

$$ab = f^2 \qquad ac = e^2 \qquad ad = ef \qquad bc = d^2.$$

This subset $R_1$ is given by four equations, nevertheless it is a two dimensional subset (because the equations are not 'independent'). In fact $R_1$ is isomorphic to $P^2(\mathbf{C})$, because a rank one conic is a line of multiplicity two and the space of lines in $P^2(\mathbf{C})$ is isomorphic to $P^2(\mathbf{C})$ itself by duality.

2. The **rank two** conics correspond to the matrices A given by the equation det A = 0 (more correctly this is the subset consisting of rank one and rank two conics) and so this is the set defined by the equation

$$abc + 2def - ad^2 - be^2 - cf^2 = 0.$$

Such a set is called a hypersurface of degree 3 in $P^5(\mathbf{C})$. The rank two conics are pairs of lines. The set of unordered pairs of lines is $P^2(\mathbf{C}) \times P^2(\mathbf{C})/(x,y) \sim (y,x)$. The rank one conics sit in this set as the diagonal elements $(x,x)$. This degree three hypersurface is homeomorphic to $P^2(\mathbf{C}) \times P^2(\mathbf{C})/\sim$

3. The **rank three** conics correspond to the complement of this degree three hypersurface.

**Polarity**

This is a concept that applies in any projective space $P^n(F)$ but we will concentrate on the case $n = 2$. As a small project, the interested reader could carry out the generalization himself.

The concept of polarity involves a fixed, non-singular conic C whose equation will be taken to be $x^tAx = 0$ where A is a $3 \times 3$ symmetric matrix. Two points $a, b \in P^2(F)$ will be called **polar** with respect to C if $a^tAb = 0$. This is a symmetric relation ($a^tAb = 0 \Rightarrow b^tAa = 0$) because A is a symmetric matrix. Note also that a point $a \in C$ is polar to itself. For a given point a, the set $\{x \mid a^tAx = 0\}$ of points polar to a forms a line called the **polar line** of a.

**Exercise** Let $\ell$ be a line in $P^2(F)$. Show that the polar lines of the points of $\ell$ are all concurrent.

Sometimes, there is a simple geometrical description of the polar line of a point:

**Proposition** Suppose that there are two tangents $\ell_1, \ell_2$ from the point a to the conic C and that they are tangent to C at $b_1, b_2$ respectively. Then the polar of the point a is the line $b_1b_2$.

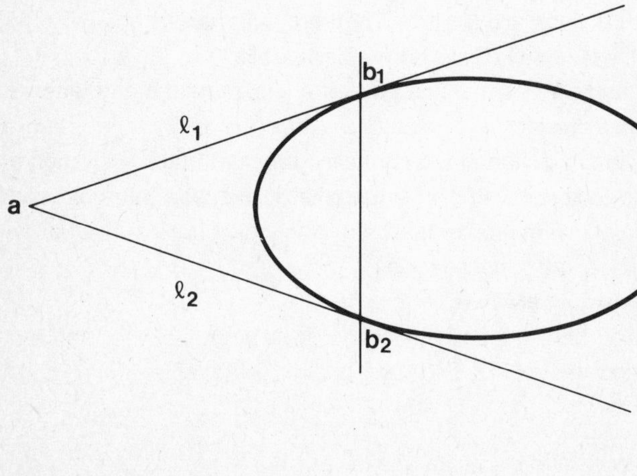

Before giving the proof, it is appropriate to explain what is meant by saying that the line $\ell$ is tangent to C.

The points of the line $\ell$ joining a and b are of the form $a + \lambda b$ ($\lambda = \infty$ gives b). So $\ell$ meets C at the points given by

$$0 = (a+\lambda b)^t A(a+\lambda b) = a^t Aa + 2\lambda b^t Aa + \lambda^2 b^t Ab.$$

This quadratic equation has either 0, 1 or 2 roots. If it has only one root, $\lambda_0$ say, then $\ell$ is said to be **tangent** to C at the point $a + \lambda_0 b$. The condition for a single root is

$$a^t Aa.b^t Ab = (b^t Aa)^2 \text{———————————} (*)$$

If $b \in C$, this condition becomes $b^t Aa = 0$ and so the equation of the tangent at b is $b^t Ax = 0$.

**Exercise** If $b \notin C$, show that a satisfies (*) if and only if a lies on one of the tangents from b to C.

**Proof of Proposition** As a lies on both tangents, we have

$$b_1{}^t Aa = b_2{}^t Aa = 0.$$

Hence, $a^t Ab_1 = a^t Ab_2 = 0$, so the polar line of a passes through $b_1$ and $b_2$, proving the Proposition.

When F = **R** and the conic C is non-empty this Proposition gives a method of constructing the polar of a point a. There are three cases depending on whether a is outside, inside or on the conic.

If a is outside the conic, draw the two tangents from a to the conic and the polar line is the line joining the two points of contact.

If a is inside the conic, draw two lines $\ell_1, \ell_2$ through a meeting the conic at $b_1$, $b_1'$ and $b_2, b_2'$ say.

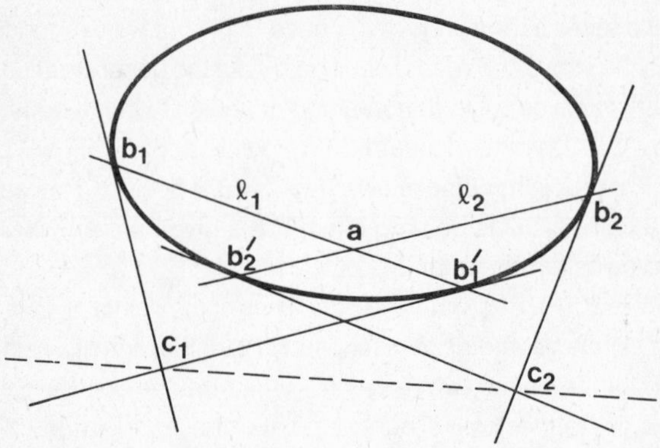


Suppose the tangents at $b_i$, $b_i'$ meet at $c_i$ ($i = 1,2$). The line $c_1c_2$ is then the polar of a.

If a is on the conic then the tangent to C at a is the polar line of a.

**Exercise** When F = **C** and $a \notin C$ prove that there is a line through a and tangent to C. Hence in this case C has no inside or outside.

**Exercise** Polarity is a generalization of orthogonality: If F = **R** and A = I, then a is polar to b if and only if corresponding vectors $\tilde{a}, \tilde{b} \in \mathbf{R}^3$ are orthogonal. In this case, the conic is empty.

An important role is played by self-polar triangles: A triangle abc is **self-polar** with respect to C if each side is the polar line of the opposite vertex.

**Proposition** Self polar triangles always exist.

**Proof** Take any point $a \notin C$ and let $\ell$ be its polar line. Take any $b \in \ell$, $b \notin C$ and let m be its polar line; m passes through a and meets $\ell$ in c. The triangle abc is self polar.

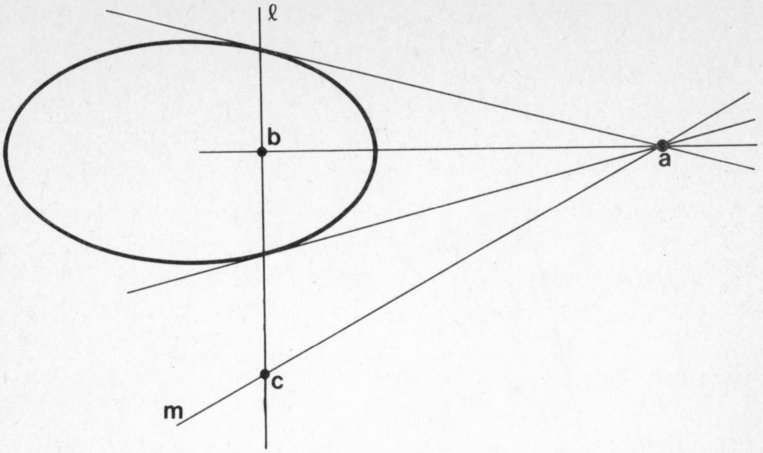


If the triangle abc is taken as the **triangle of reference**, that is, one chooses a co-ordinate system (or basis) so that a is $e_1 = [1:0:0]$, b is $e_2 = [0:1:0]$ and c is $e_3 = [0:0:1]$ then the condition that abc is self polar with respect to $x^tAx = 0$ is the same as the condition that A is diagonal: Let $A = (\alpha_{ij})$ then $e_i{}^tAe_j = \alpha_{ij}$ and if abc is self polar $e_i{}^tAe_j = 0$ if $i \neq j$ so A is diagonal. The converse is similar.

There are very many self-polar triangles, the space of all (ordered) triangles in $P^2(F)$ is a six-dimensional space (in fact $(P^2 \times P^2 \times P^2)$) and from the above proof one sees that the space of self polar triangles is three-dimensional. If two three-dimensional spaces lie in a six dimensional space then one expects them to meet (just as two general three dimensional affine spaces in $\mathbf{R}^6$ meet). On these rather general grounds one would expect two conics to have a common self polar triangle and this is in fact the case:

**Proposition** Suppose $C_1$, $C_2$ are two conics in $P^2(F)$ that meet in four distinct points. Then, there is a (unique) triangle which is self-polar with respect to both $C_1$ and $C_2$.

**Exercise** If $C_1$, $C_2$ are distinct, non-singular conics, show that $C_1 \cap C_2$ has at most four points.

As a preliminary we will study how the conic, a point and its polar meet a line. But first we need a new notion.

**Definition** Two pairs of collinear points, a,b and $\alpha,\beta$ are **harmonic** if $(a,b;\alpha,\beta) = -1$.

**Lemma** The pairs a,b; $\alpha,\beta$ are harmonic if and only if $(a,b;\alpha,\beta) = (b,a;\alpha,\beta)$.

**Proof** If $\lambda = (a,b;\alpha,\beta)$ then $\lambda^{-1} = (b,a;\alpha,\beta)$. If the four points a, b, $\alpha$, $\beta$ are distinct then $\lambda \neq 1$. So $(a,b;\alpha,\beta) = (b,a;\alpha,\beta) \Leftrightarrow \lambda^2 = 1 \Leftrightarrow \lambda = -1$.

Harmonic pairs arise naturally from complete quadrilaterals:

**Proposition** Let a, b, c, d be four points in $P^2(F)$. Let $p = (ad) \cap (bc)$, $q = (ac) \cap (bd)$ and $r = (ab) \cap (cd)$. Let $s = (qr) \cap (ad)$ and $t = (qr) \cap (bc)$. Then $(q,r;s,t)$ is harmonic.

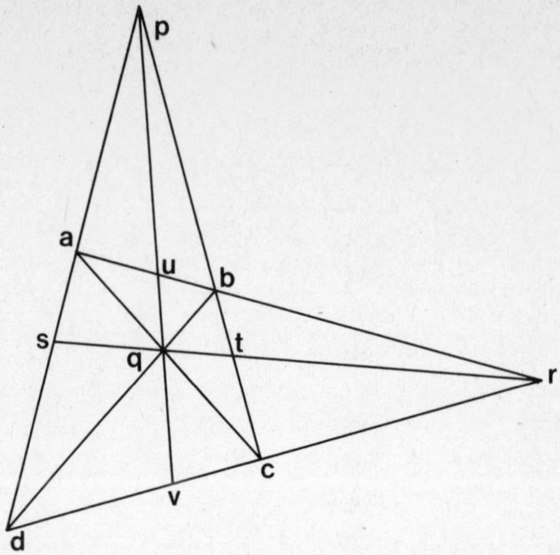


**Proof** Let $u = (pq) \cap (ab)$ and $v = (pq) \cap (cd)$. Projecting from p one gets

$$(q,r;s,t) = (v,r;d,c)$$

and by projecting from q one gets

$$(v,r;d,c) = (u,r;b,a).$$

Projecting once again from p gives

$$(u,r;b,a) = (q,r;t,s).$$

The result now follows from the lemma.

**Exercise** If $a, b, c \in \mathbf{R}$ then the pairs a, b and c, $\infty$ are harmonic if and only if c is the mid point of ab.

We now return to conics:

**Lemma** Let C be a non-degerate conic and let $\ell$ be the polar of the point a. If m is a line through a meeting $\ell$ at b and meeting C in $\alpha$, $\beta$ then the pairs a,b; $\alpha,\beta$ are harmonic. Conversely if $(a,b;\alpha,\beta) = -1$ then b is the point of intersection of $\ell$ and m.

**Proof**

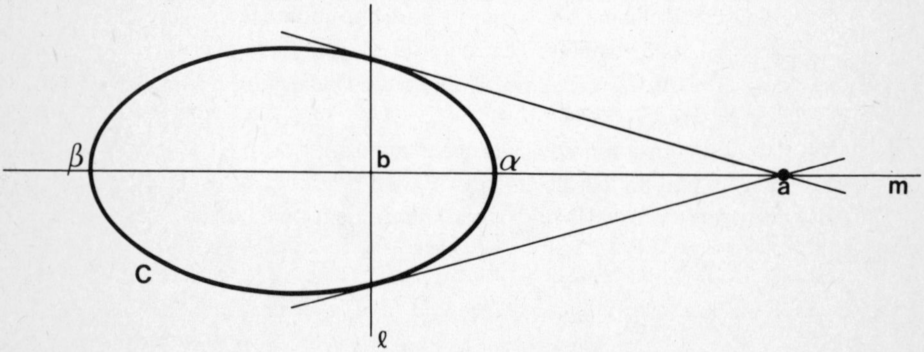


Choose co-ordinates so that $\alpha$ is [1:0:0] and $\beta$ is [0:1:0] so m is the line $z = 0$. The equation of C is therefore $xy + z\lambda = 0$ where $\lambda$ is some linear expression in x, y, z. Let a be [1:s:0] and b be [1:t:0] then a and b are polar if and only if $s + t = 0$. But $(a,b;\alpha,\beta) = s/t$ and so the result follows.

Now we can return to the proof of the existence of a self polar triangle.

**Proof** Let a, b, c, d be the four points where the conics meet, construct the complete quadrilateral from them as on page 72. The triangle p, q, r is self-polar: It is enough to prove that pq is the polar of r. The pairs a,b;r,u are harmonic so, as a, b $\in$ C, u must lie on the polar of r. Similarly v lies on the polar of r, proving the result.

Combining the relationship between diagonalization and self polarity with the existence of a common self polar triangle one sees that it is almost always possible to diagonalise two quadratic forms simultaneously. A sufficient condition being that the two conics must meet in four points.

**Exercise** Prove that two quadratic forms cannot be diagonalized simultaneously if the corresponding conics touch.

Given a triangle, there are many conics with respect to which it is self polar:

**Proposition** Let abc be a triangle and p $\in$ (ab), q $\in$ (ac) then there is a unique conic C such that abc is self polar and such that p, q both lie on C.

**Proof** We can take abc to be the triangle of reference. Then abc is self polar with respect to a conic C if and only if the equation of C is

$$\lambda x^2 + \mu y^2 + \nu z^2 = 0.$$

The conditions that p, q lie on C uniquely determine $\lambda$, $\mu$, $\nu$ up to a constant multiple.

Now, we discuss the famous theorem of Pascal. To prove it we need to consider the cross-ratios of points on a conic.

**Lemma** Let p, q be two distinct points on a non-singular conic C and let a, b, c, d be four other distinct points of C, then (pa,pb;pc,pd) = (qa,qb;qc,qd).

**Proof** Let pa, qa, pb, qb have equations $\ell_a x = 0$, $m_a x = 0$, $\ell_b x = 0$, $m_b x = 0$. Any line through p has equation $(\lambda\ell_a+\mu\ell_b)x = 0$ and similarly for the lines through q. Without loss of generality, we can assume that the equations of pc, qc are $(\ell_a-\ell_b)x = 0$ and $(m_a-m_b)x = 0$.

The conic C passes through the four points of intersection of the two line pairs (pa) $\cup$ (qb) and (pb) (qa). The equations of these line pairs are

$$(\ell_a x)(m_b x) = 0 \qquad \text{and} \qquad (\ell_b x)(m_a x) = 0$$

The matrix of the conic C rnust be a combination of the matrices of these two line pairs as they all have four points in common. So the equation of C is of the form

$$s(\ell_a x)(m_a x) + t(\ell_b x)(m_b x) = 0.$$

But the equation of C is also a combination of those of (pa) $\cup$ (qc) and (pc) $\cup$ (qa) so s = –t = 1. Hence the equation of C is

$$(\ell_a x)(m_b x) = (\ell_b x)(m_a x) \text{———————} (*)$$

A calculation now shows that

$$(pa,pb;pc,pd) = \ell_a d/\ell_b d$$

and

$$(qa,qb;qc,qd) = m_a d/m_b d.$$

But d $\in$ C so by (*) the two cross-ratios are equal.

**Corollary** Given four points a, b, c, d on a conic C one can define their cross-ratio (a,b;c,d) to be (pa,pb;pc,pd) for any p $\in$ C.

**Pascal's Theorem** Let a,b,c;a′,b′,c′ be six distinct points on a non-singular conic. Then the three points

$$p = (bc') \cap (b'c),\; q = (ac') \cap (a'c) \text{ and } r = (ab') \cap (a'b)$$

are collinear.

**Proof**

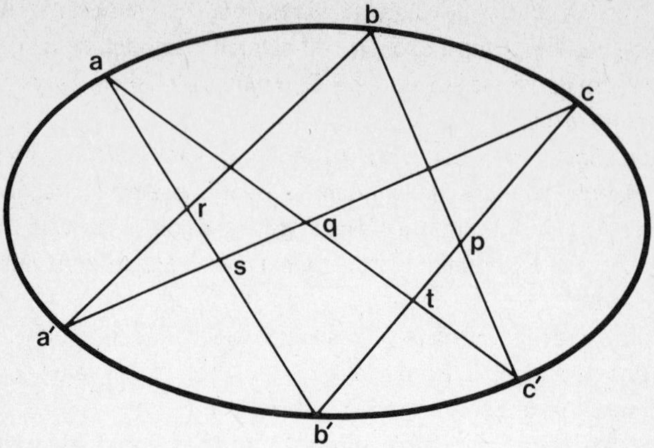


Let s = (ab′) ∩ (a′c) and t = (ac′) ∩ (b′c).

By projecting from a′ and using the above corollary one gets (a,s;r,b′) = (a,c;b,b′).

Similarly, using c′ one gets (a,c;b,b′) = (t,c;p,b′).

So (qa,qs;qr,qb′) = (qt,qc;qp,qb′).

But qa = qt and qs = qc so qr = qp, hence p, q, r are collinear.

**Exercise** Use the same method to prove Pappus's theorem: If a, b, c are collinear and a′, b′, c′ are collinear, prove that p, q, r are collinear where p = (bc′) ∩ (b′c), q = (ac′) ∩ (a′c) and r = (ab′) ∩ (a′b).

Note that Pappus's theorem can be regarded as a degenerate case of Pascal's theorem – when the conic is singular.

# Problems

1. Show that the equation of the line joining the two distinct points $[a_1:b_1:c_1]$ and $[a_2:b_2:c_2]$ in $P^2$ is

$$\det\begin{bmatrix} x & y & z \\ a_1 & b_1 & c_1 \\ a_2 & b_2 & c_2 \end{bmatrix} = 0$$

2. Let V be an m-dimensional subspace of $\mathbf{R}^n$ and let $L_1, L_2$ be two (n-m)-dimensional subspaces of $\mathbf{R}^n$ such that $V \cap L_1 = V \cap L_2 = \{0\}$. Define a mapping $f: L_1 \to L_2$ as follows: for $x \in L_1$, consider the affine space $x + V$, it meets $L_2$ in a single point y, define $f(x) = y$. Prove that f is a linear map.

3. Show that the maps $f: S^2 \to \mathbf{R}^4$, $g: D^2 \to \mathbf{C}^2$, defined by

$$f(x,y,z) = (x^2-y^2, xy, yz, zx),$$

and
$$g(re^{i\theta}) = (re^{2i\theta}, r(1-r)e^{i\theta}),$$

induce continuous, injective maps of $P^2$ into $\mathbf{R}^4$.

4. The faces of a regular octahedron can be coloured black and white so that no two adjacent faces are of the same colour. Consider the object formed by the union of the black faces and the three diagonal squares of the octahedron. This is called the heptahedron. Show (by cutting and pasting) that there is a continuous map from $P^2$ to the heptahedron. So the heptahedron is a model for $P^2$ in $\mathbf{R}^3$.

5. Formulate the dual of Pappus's theorem. You should dualise some other theorems of projective geometry.

6. Four points $a, b, c, d \in P^2$, no three of which are collinear, determine a complete quadrangle. It has six lines, which have three other points in common. Draw this configuration. What is its dual?

7. Let $\pi_1, \pi_2$ be two copies of $P^{n-1}(F)$ in $P^n(F)$ and let $p \notin \pi_1 \cup \pi_2$. Prove that the projection of $\pi_1$ onto $\pi_2$ from p induces a projectivity of $P^{n-1}(F)$. (Sometimes such a map is called a perspectivity with centre p.)

   Show that any projectivity: $P^1(F) \to P^1(F)$ can be expressed as the composite of two perspectivities.

8. By considering the Jordan normal form, show that there are six essentially different types of elements in $PGL(3,\mathbf{C})$. Can they be distinguished by their sets of fixed points?

9. Check that the obvious homomorphism $SL(2,\mathbf{R}) \to PGL(2,\mathbf{R})$ has kernel of order two, but that the two spaces are homeomorphic.

10. Let $a_0, a_1, \ldots, a_n$ be (n+1) points in $P^n(F)$ that do not lie in any hyperplane. Prove that the subgroup of $PGL(n+1,F)$ that consists of those f such that $fa_i = a_i$ for $0 \leqslant i \leqslant n$ is isomorphic to $(F^*)^n$. [$F^* = F \setminus \{0\}$.]

11. Prove that the map $f: \mathbf{C} \to \mathbf{C}^*$ defined by $f(z) = \exp(2\pi iz)$ is a homomorphism whose kernel is $\mathbf{Z}$. Deduce that $\mathbf{C}^* \cong \mathbf{C}/\mathbf{Z}$.

12. Given (n+2) points $p_0, p_1, \ldots, p_{n+1}$ in $P^n(F)$, no (n+1) of which lie in a hyperplane, and another such set $q_0, q_1, \ldots, q_{n+1}$, prove that there is a unique $f \in PGL(n+1,F)$ such that $fp_i = q_i$ for all $0 \leqslant i \leqslant n+1$.

If $p_0, p_1, \ldots, p_4$ are distinct points in $P^2(F)$ and $p_0, p_1, p_2$ are not collinear but both the sets $p_0, p_1, p_3$ and $p_0, p_2, p_4$ are collinear, prove that there is a unique $f \in PGL(3,F)$ sending the p's onto any set $q_0, q_1, \ldots, q_4$ that has the same properties.

13. Find a monomorphism $S_{n+1} \to PGL(n,\mathbf{C})$ where $S_{n+1}$ is the symmetric group on n + 1 letters. Describe the image of this homomorphism geometrically.

14.* Show that the image of $SU(2) \to GL(2,\mathbf{C}) \to PGL(2,\mathbf{C})$ is isomorphic to SO(3) and that the action of $f \in PGL(2,\mathbf{C})$ on $P^1(\mathbf{C})$ is the same as that of the corresponding element of SO(3) on $S^2$. (Corresponding under stereographic projection.) In this way the rotation groups of the Platonic solids can be identified with groups of Möbius transformations. Verify the following.
   i) The tetrahedral group is generated by the Möbius transformations $T_1(z) = (z+i)/(z-i)$, $T_2(z) = -1/z$.
   ii) The octahedral group is generated by $T_1$, $T_2$ and $T_3(z) = iz$.
   iii) The icosahedral group is generated by
   $$\omega z,\ [(\omega^4-\omega)z + (\omega^2-\omega^3)]/[(\omega^2-\omega^3)z + (\omega-\omega^4)],\ -1/z$$
   where $\omega$ is a primitive fifth root of unity.
   Find Möbius transformations that generate the dihedral groups.

15.* Show that, for each $k > 0$, the group $GL(n,\mathbf{C})$ acts on the space of homogeneous polynomials of degree k in n variables. Verify that the three polynomials
$$x^4 \pm 2i\sqrt{3}\,x^2y^2 + y^4, \qquad xy(x^4-y^4)$$
are each invariant (up to a constant factor) under the action of the tetrahedral group. Is their stabilizer equal to the tetrahedral group? How are the zeros of these polynomials related to the tetrahedron?

16. For $f, g \in PGL(2,\mathbf{C})$, prove that $fg = gf$ if and only if either
   i) f or g is the identity
   ii) f and g have the same fixed point sets, or
   iii) $f^2 = g^2 = 1$ with fixed points $x_1, x_2; y_1, y_2$ that satisfy $(x_1,x_2;y_1,y_2) = -1$.

17. If $(ab,cd) = \lambda$, calculate the cross-ratios of the various permutations of a, b, c, d in terms of $\lambda$ (for example, $(ad,bc) = (\lambda-1)/\lambda$). Hence find a non-trivial homomorphism $S_4 \to S_3$ between the symmetric groups.
[Hint: consider the action of $S_4$ on the set $\{2,2^{-1},-1\}$ obtained using $\lambda = -1$.]
   Use a similar method to construct a non-trivial homomorphism $S_4 \to \{\pm 1\}$.
   Note that for $n > 4$, the only non-trivial homomorphism defined on $S_n$ is the sign homomorphism.
   The non-trivial homomorphism $S_4 \to S_3$ can be constructed more geometrically as follows. Consider the complete quadrilateral drawn on page 72. Regard $S_4$ as the group of bijections of the set $\{a,b,c,d\}$ and $S_3$ as the group of bijections of $\{p,q,r\}$. A permutation of $\{a,b,c,d\}$ induces one of $\{p,q,r\}$.

18. Show that, for any three distinct points $a, b, c \in P^1(F)$, the map $x \to (ab,cx)$ defines a bijection
$$P^1(F) \setminus \{a\} \to F.$$

19. Let $f: P^1(\mathbf{R}) \to P^1(\mathbf{R})$ be a map such that $(ab,cd) = (fafb,fcfd)$ whenever a, b, c, d are distinct. Show that $f \in PGL(2,\mathbf{R})$. Is the corresponding result true for $P^1(\mathbf{C})$?

20. The set of lines through a fixed point $p \in \mathbf{R}^2$ can be identified with $P^1(\mathbf{R})$. If $\ell$, m are two lines meeting at p, and $b_1$, $b_2$ are the bisectors of the angles between them, show that $(\ell b_1, m b_2) = -1$.

21. Set up projective spaces over the quaternions.
Find out about the Cayley numbers (an eight dimensional algebra over **R**). In this case one can define the projective plane but not a higher dimensional projective space. Investigate.

22. On a sphere of unit radius consider a spherical triangle with angles and sides as shown, each angle being less than $\pi$. Prove that
i) $\cos a = \cos b \cos c + \sin b \sin c \cos\alpha$
ii) $\sin a \sin\beta = \sin b \sin\alpha$.

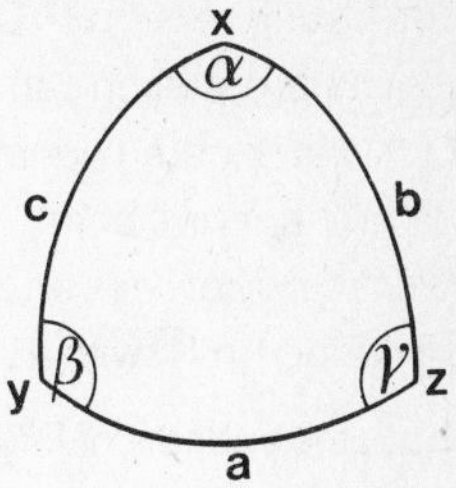


23. Show that there is a spherical triangle with angles $0 < \alpha \leq \beta \leq \gamma$ if and only if
$$\beta + \gamma < \pi + \alpha.$$
For which integers p, q, r is there a spherical triangle whose angles are $\pi/p$, $\pi/q$ and $\pi/r$?

24. Given the area and base of a spherical triangle, find the locus of the third vertex.

25. A circle in the elliptic plane is a set $\{y | d(y,x) = r\}$ for a fixed $x \in P^2$ and $r > 0$. By drawing the corresponding sets in $S^2$, find two circles in the elliptic plane that meet in four points.

26. Sketch the curve $y^2 = x^3 + x^2$ in $\mathbf{R}^2$. What is the equation of the corresponding curve in terms of homogeneous coordinates for $P^2(\mathbf{R})$? Sketch that part of the curve that lies in the part of $P^2(\mathbf{R})$ given by $x = 1$, indicate in this sketch those points of the curve that lie in the original $\mathbf{R}^2$.

27. Let C be a non-singular conic in $P^2(\mathbf{C})$. Prove that there are linearly independent homogeneous quadratics $f_i(\lambda,\mu)$, $(i = 1, 2, 3)$ such that each point of C is of the form
$$[f_1(\lambda,\mu):f_2(\lambda,\mu):f_3(\lambda,\mu)].$$

28. Given five points in $P^2(\mathbf{F})$ no three of which are collinear, prove that there is a unique non-singular conic passing through them. What happens when three of the points are collinear? Are there any special cases?

29.* Let V be a vector space over the field F and $\varphi: V \to F^2$ be a linear map of rank 2 with kernel K. Show that there is an induced map
$$\varphi: \mathbf{P}(V) \setminus \mathbf{P}(K) \to P^1(F).$$
If a, b, c, d are collinear points in $\mathbf{P}(V) \setminus \mathbf{P}(K)$ show that $(ab,cd) = (\varphi a \varphi b, \varphi c \varphi d)$. By constructing suitable maps $\varphi$, show that if a, b, c, d are points on a non-singular conic C in $P^2(F)$, then the cross-ratio (ab,cd) can be defined as the cross-ratio of the lines joining any point $p \in C$ to a, b, c, d.

30.* Consider the space X of all lines in $P^3(F)$. If $\ell$ is the line joining $[a_0:a_1:a_2:a_3]$ and $[b_0:b_1:b_2:b_3]$, let $P_\ell$ be the $4 \times 4$ skew-symmetric matrix with entries
$$\det\begin{bmatrix} a_i & a_j \\ b_i & b_j \end{bmatrix} = p_{ij}.$$
Show that these entries satisfy
$$p_{01}p_{23} + p_{02}p_{31} + p_{03}p_{12} = 0 \text{ ——— } (*)$$

Show that $\ell \to P_\ell$ defines a map $X \to Q$ where Q is the quadric in $P^5(F)$ whose equation is (*). Conversely show that each point Q corresponds to a unique line in $P^3(F)$, so that X is isomorphic to Q.

31. Let Q be a non-singular quadric in $P^3(\mathbf{R})$, that is $Q = \{x|x^tAx = 0\}$ where A is a non-singular, symmetric real $4 \times 4$ matrix. By putting A into diagonal form, show that if Q is not empty then it is homomorphic either to a torus $T^2$ or to a sphere $S^2$.

32.* Consider a surface X in $P^3(\mathbf{C})$ defined by a non-singular quadratic form. Show that by using a projectivity its equation can always be put in the form $x_0x_3 = x_1x_2$. Deduce that through every point x of X there are two lines $\ell_1, \ell_2 \subset X$. If $m_1$, $m_2$ are the lines through a different point y of X, show that one of $m_1$, $m_2$ meets $\ell_1$ in one point and the other does not meet $\ell_1$. This gives two families of lines on X.

Find a surface in $\mathbf{R}^3$ with a similar property.

33. Investigate the subsets of $P^5(\mathbf{R})$ corresponding to the five types of real conics.

34. (For those with a knowledge of the calculus of several variables.) Consider the map $\delta: S^5 \to \mathbf{R}$ defined by

$$\delta(a,b,c,d,e,f) = abc + 2def - ad^2 - be^2 - cf^2.$$

At which points of $\delta^{-1}(0)$ is $\delta$ regular? Show that the image of $\delta^{-1}(0)$ in $P^5$ is $P^2 \times P^2/\sim$ where $(x,y) \sim (y,x)$. Is $P^2 \times P^2/\sim$ a manifold?

35. Consider the family of conics $x^2 + y^2 = tz^2$ in $P^2(\mathbf{R})$ and $P^2(\mathbf{C})$, as t varies. If $t \neq 0, \infty$ the conic is non-singular. Describe how it degenerates to a singular conic as $t \to 0$ and $t \to \infty$.

Repeat with the family $x^2 + ty^2 = z^2$.

36. What is the intersection of the Veronese surface (page 69) with a linear $P^3(F) \subset P^5(F)$?

37. Consider four lines in $P^2$, that meet in six points. Label the six points as a, $a^1$, b, $b^1$, c, $c^1$ in such a way that a and $a^1$ do not lie on any one line and similarly for b, $b^1$ and c, $c^1$. Let C be a non-singular conic such that a, $a^1$ and b, $b^1$ are polar pairs with respect to C. Prove that c, $c^1$ are also polar pairs.

Hence or otherwise show that a triangle and its polar are in perspective, that is the lines $aa^1$, $bb^1$ and $cc^1$ are concurrent.

38. Define a map $I: \mathbf{R}^2 \setminus \{0\} \to \mathbf{R}^2 \setminus \{0\}$ by $|0Ix|.|0x| = 1$ and Ix is on the half line 0x. Show that I maps circles to circles except that circles through 0 are mapped to lines and vice versa. Is I conformal?

The point 0 is 'sent' to infinity under I. Find a space X containing $\mathbf{R}^2$ and a map $\bar{I}: X \to X$ such that $\bar{I}|\mathbf{R}^2 \setminus \{0\}$ is I. Can you write down a formula for $\bar{I}$?

# Part III

# Hyperbolic Geometry

## The Parallel Axiom

Euclid gave certain axioms that were meant to characterize plane geometry. For twenty centuries much controversy surrounded one of these axioms – the parallel axiom. It was felt that this axiom was not as basic as the others and considerable efforts were put into attempts to deduce it from the other axioms. Before the nineteenth century the most successful attempt seems to have been made by Saccheri in 1773. He produced a thorough investigation of the parallel axiom, in fact he claimed to have deduced the parallel axiom from the others, but his otherwise excellent piece of work had one flaw in it. At the beginning of the nineteenth century Gauss, in work that he did not publicise, showed how a large body of theorems could be deduced from a variant of the parallel axiom, thus discovering hyperbolic geometry. Because Gauss did not publish his work, this discovery is usually associated with the names of Bolyai and Lobachevski who published the same discoveries in 1832 and 1836 respectively. Despite the fact that Euclid's parallel axiom was shown to be necessary several defects were found, during the nineteenth century, both in the axioms and logic used by Euclid. For a discussion of Euclid's axioms and their defects, the reader may consult M.J. Greenberg's book "Euclidean and Non-euclidean geometry".

The parallel axiom of the Euclidean plane may be stated as: given a line $\ell$ and a point $p \notin \ell$ there is a unique line m with $p \in m$ and $\ell \cap m = \phi$.

The corresponding fact for the projective plane is: given a line $\ell$ and a point $p \notin \ell$ there is no line m with $p \in m$ and $\ell \cap m = \phi$.

The only alternative (assuming homogeneity) is: given a line $\ell$ and a point $p \notin \ell$ there is more than one line m such that $p \in m$ and $\ell \cap m = \phi$.

It is this third possibility that holds in hyperbolic geometry. At first sight it seems hard to imagine such a geometry. The first simple 'model' of such a geometry was discovered by Beltrami in 1868.

## The Beltrami (or projective) Model

Let C be a non-singular conic in $P^2$; as we have seen C divides $P^2$ into two parts – an inside (homeomorphic to $\mathbf{R}^2$) and an outside (homeomorphic to the Möbius band M). The inside will be denoted by H and is the space of the geometry, it remains to describe the lines of the geometry. A line in H is a chord of the conic C, that is $H \cap \ell$ where $\ell$ is a line in $P^2$. The conic C is called the **absolute conic** of H. To obtain a picture let the conic C be an ellipse in $\mathbf{R}^2$.

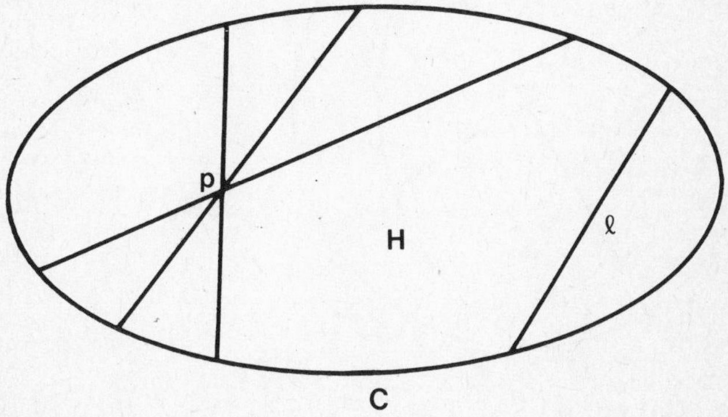

For a point $p \in H$ with $p \notin \ell$, there are many lines of the geometry not meeting $\ell$. So, the third possibility for the parallel axiom holds.

Two lines are **parallel** if they meet on C, they are **ultraparalel** if they meet outside C.

An obvious group that sends lines of H to lines of H is the group

$$G_C = \{f \in PGL(3,\mathbf{R}) | fC = C\}.$$

That is, it consists of the projectivities of $P^2$ that send the absolute conic C to itself. Such a projectivity must send the complement of C to itself. This complement has two components, and they are not homeomorphic to each other. So, a projectivity f that sends C to itself must send H to itself.

It turns out that there is a metric on H, such that the group $G_C$ is the group of direct isometries. It is not at all clear that there should be such a metric. The next section attempts to give general reasons why there might be such a metric, and later the metric is constructed.

The reader who only wants to see how the metric is constructed may skip the next section.

## Groups of plane isometries

We analyse the 'size' of the groups of isometries of the Euclidean and elliptic planes. Then we show that $G_C$ has the same size. We measure the size by answering the following question. Given two points p, q of the geometry and two lines $\ell$, m such that $p \in \ell$ and $q \in m$, how many isometries f are there such that $fp = q$ and $f\ell = m$?

a) **The Euclidean plane**

There is a translation that sends p to q and composing with a rotation about q shows that there is at least one isometry f with the required property. There are exactly four such, all obtained from each other by first composing with reflection in $\ell$, reflection in a line perpendicular to $\ell$, and the half turn about p which is the composite of the two reflections.

b) **The elliptic plane**

The isometries of the elliptic plane form the group PO(3), and this is isomorphic to SO(3) the group of rotations of $S^2$. So we consider SO(3) acting on $S^2$. Given a point $p \in S^2$ and a great circle $\ell$ containing p, it is clear that there is a rotation sending the pair p, $\ell$ to any other such pair q, m. It remains to show that there are exactly four rotations that take p, $\ell$ to itself or to $-p$, $\ell$, because these are the only points that are identified under the projection from $S^2$ to the elliptic plane $P^2$. Consider the diagram

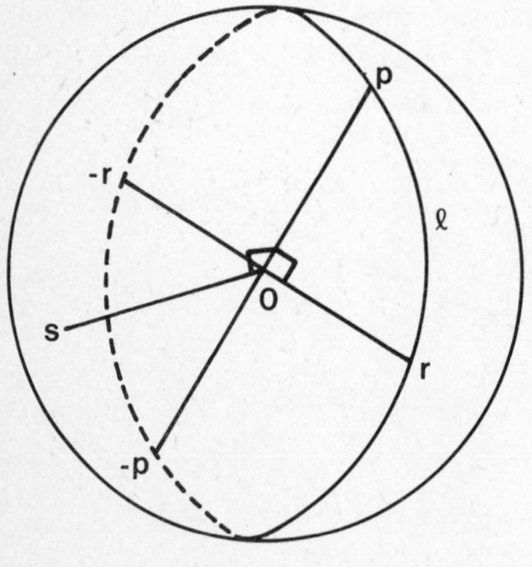


$p, r \in \ell$

$Op \perp Or$

$Or \perp Os$

$Op \perp Os$

The only rotations that send p to p and $\ell$ to $\ell$ are the identity and the half-turn about Op. The only rotations that send p to –p and $\ell$ to $\ell$ are the half-turns about Or and Os. These four rotations give four distinct elements of PO(3) because SO(3) → PO(3) is an isomorphism.

c) **The Beltrami model**

To prove the result in this case we need two results of projective geometry.

1) Given two triangles abc and $a^1b^1c^1$ in $P^2(\mathbf{R})$ and points $x \in ab$, $y \in ac$, $x^1 \in a^1b^1$, $y^1 \in a^1c^1$, there is a unique $f \in PGL(2,\mathbf{R})$ sending x, y, a, b, c to $x^1$, $y^1$, $a^1$, $b^1$, $c^1$ respectively. (Problem 12 page 76.)

2) Given a triangle abc and points $x \in ab$, $y \in ac$, there is a unique conic C such that the triangle abc is self-polar with respect to C and such that $x \in C$, $y \in C$. (see page 73).

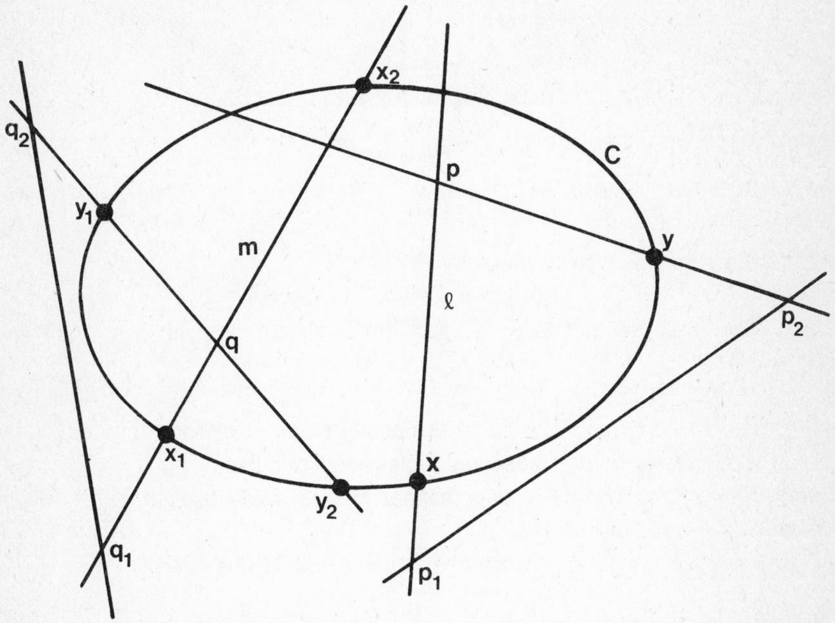


Let C be the absolute conic of the model. Given $p \in \ell$, there is a unique self-polar triangle $pp_1p_2$ such that $p_1 \in \ell$: $p_2$ is the polar of $\ell$ and $p_1$ is the point where the polar of p meets $\ell$. Similarly let $qq_1q_2$ be self-polar with $q_1 \in m$.

Choose x, y such that $x \in \ell \cap C$, $y \in pp_2 \cap C$. Let $\{x_1, x_2\} = m \cap C$ and $\{y_1, y_2\} = qq_2 \cap C$.

By 1) above there is a unique projectivity $f_{ij}$ sending p, $p_1$, $p_2$, x, y to q, $q_1$, $q_2$, $x_i$, $y_j$. By construction, $x_i \in m \cap C$, $y_j \in qq_2 \cap C$ and $qq_1q_2$ is self-polar with respect to C. The conic fC has the same properties, but by 2) above, there is a unique such conic, so fC = C. Hence there are four elements, $f_{11}$, $f_{12}$, $f_{21}$ and $f_{22}$ of $G_C$ that send p,$\ell$ to q,m.

We will construct the metric on the Beltrami model but first we will construct another model, the Poincaré model. To motivate this we need to study stereographic projection.

**Stereographic Projection**

This is the map that identifies the two dimensional sphere $S^2 \subset \mathbf{R}^3$ with $\mathbf{P}^1(\mathbf{C})$. Here we will regard it as a map $S: S^2 \setminus \{n\} \to \mathbf{C}$, where n is the north pole of the sphere. There are several variants of the definition. We choose the one where $\mathbf{C}$ is the plane through the equator of $S^2$ (any parallel plane would do almost as well). Define $Sp = p'$ if n, p and $p'$ are collinear.

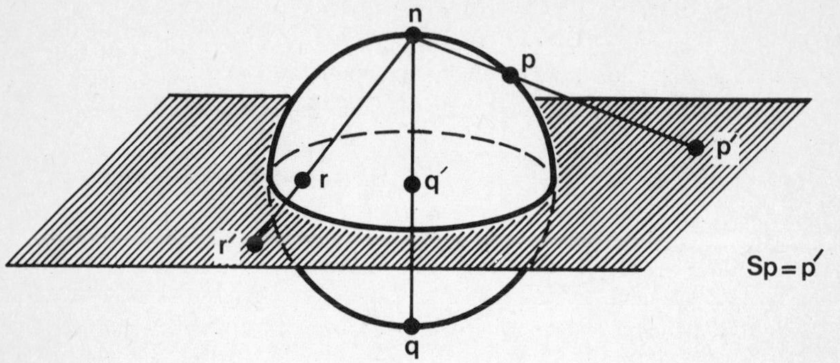


This map can be described in co-ordinates by

$$S(x,y,z) = (x+iy)/(1-z) \text{ for } z \neq 1.$$

**Exercise** Show that two points $p, q \in S^2$ are antipodal, that is $p = -q$ if and only if $S(p)\overline{S(q)} = -1$.

Stereographic projection S has two important properties that we will need:

1. Conformality, in other words it preserves angles between curves.
2. Circle preserving, a circle on $S^2$ gets mapped to a circle in $\mathbf{C}$ except that a circle through n gets mapped to a straight line in $\mathbf{C}$.

The main features of the proofs of these properties will now be given.

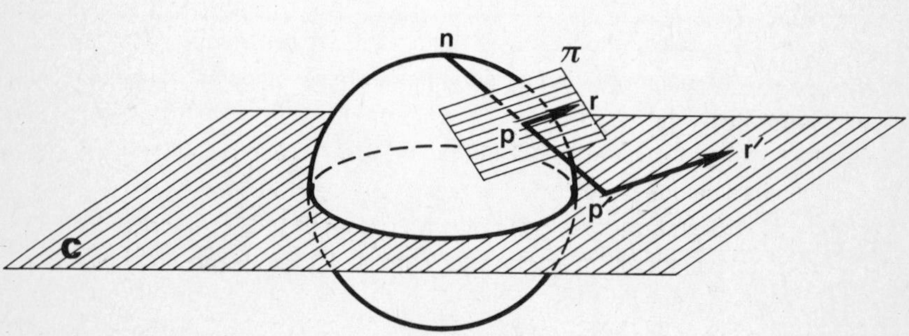


1.

Let $\pi$ be the tangent plane to $S^2$ at p and pr be a line in $\pi$. Let $p'r'$ be the image of pr under S. The plane $\pi$ makes the same angle with $\mathbf{C}$ as it does with the tangent plane at n.

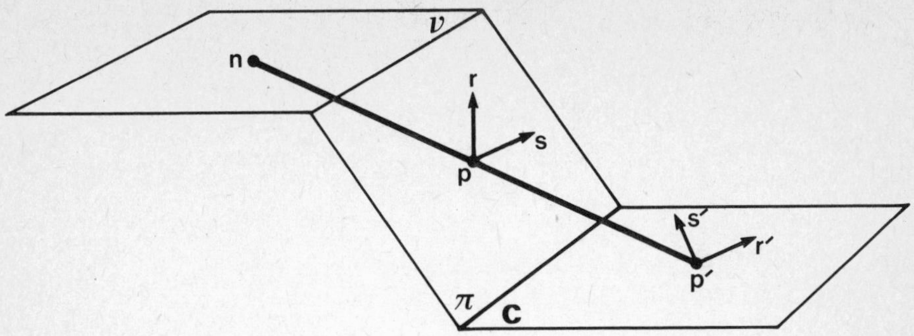


The line np′ is perpendicular to the line $\pi \cap \mathbf{C}$, because it is perpendicular to the intersection of the tangent planes $\nu$ and $\pi$ and $\nu$ is parallel to **C**. Hence the angles rps and r′p′s′ are equal – as was required to be proved.

2. Consider a circle C on $S^2$ not through n, the cone that is tangential to the sphere along C has its vertex at x say, we will show that the stereographic projection of C is a circle whose centre is the projection y of x onto the plane **C**.

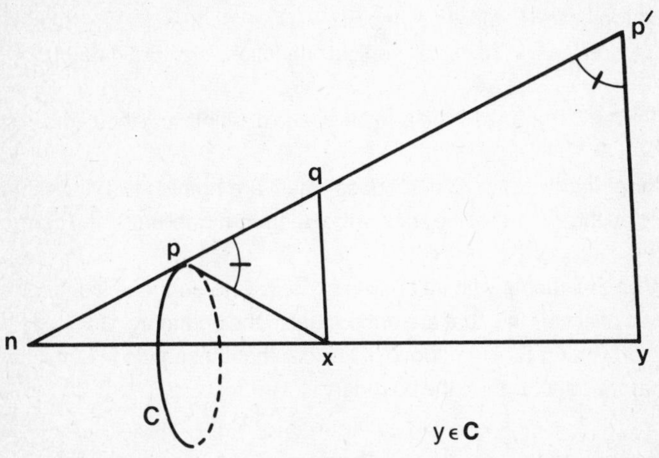


This diagram is in the plane of n, p and x; xq is drawn parallel to p′y. We will show that the length p′y is independent of where p is on the circle C. Using the symmetry of pp′ with respect to the planes $\pi$ and **C**, it is clear that the angles qpx and qp′y are equal. So $|px| = |qx|$. Therefore $|p'y| = |ny|.|qx|/|nx| = |ny|.|px|/|nx|$. As $|px|$ is constant for $p \in C$ we see that $|p'y|$ is constant as required.

The case where the circle C goes through n is much simpler.

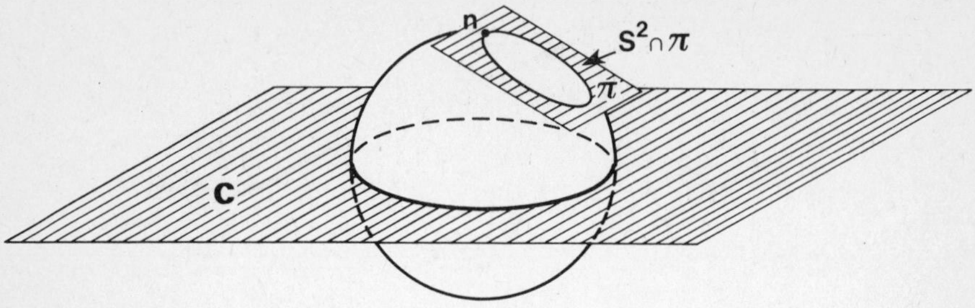


Clearly the projection of the circle $S^2 \cap \pi$ is the line $\mathbf{C} \cap \pi$.

**The Poincaré Model**

We will define this model as the image of the Beltrami model under certain projections. Let D be the open disc of unit radius in **C**,

$$D = \{z \in \mathbf{C} \mid |z| < 1\}$$

and L be the lower hemisphere of $S^2$

$$L = \{(x,y,z) \in S^2 \mid z < 0\}.$$

Clearly stereographic projection S defines a map $S: L \to D$.
There is another projection $V: L \to D$ given by vertical projection, in co-ordinates,

$$V(x,y,z) = x + iy.$$

The Beltrami model can be taken to have D for its space of points and the chords of the boundary circle to be its lines.

The **Poincaré model** has D for its space but its 'lines' are nothing like ordinary straight lines. Its lines are defined to be the images of the lines of the Beltrami model under the composite

$$SV^{-1}: D \to L \to D,$$

where we take the Beltrami model with its boundary conic C to be the unit circle. Notice that $V^{-1}$ takes lines in D to arcs of circles in L that are orthogonal to the boundary. Using the two properties of the stereographic projection S given above, it is clear that these arcs in L are mapped by S to arcs of circles in D that are orthogonal to the boundary.

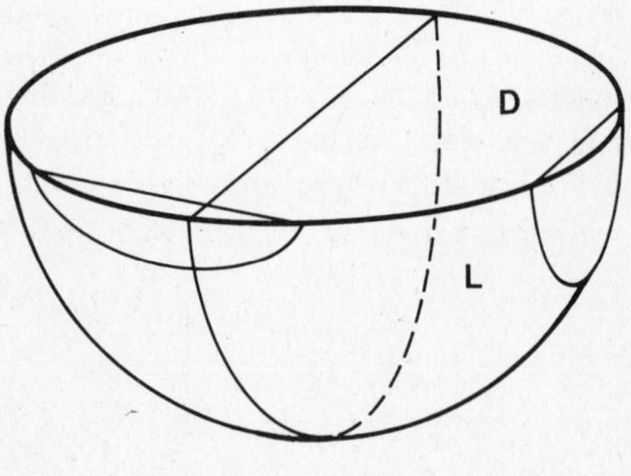

Examples of lines in the Poincaré model are:

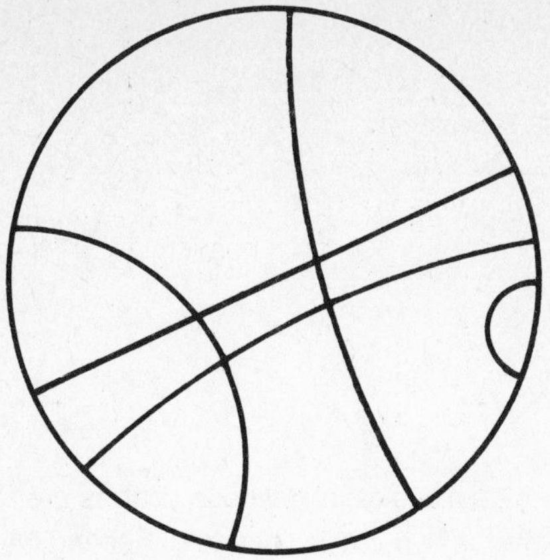

One could define the Poincaré model directly in this way: its space is D and the lines are arcs of circles perpendicular to the boundary. For the Poincaré model it is not quite obvious even that there is a unique line through any two points, although of course one knows it is true because it is true in the Beltrami model. The information can be transferred from one to the other using the map $SV^{-1}$ and its inverse $VS^{-1}$.

The reason for introducing the Poincaré model is that it is very important in mathematics. It was Poincaré's discovery of this model that brought hyperbolic geometry from a mere curiosity to an object playing a central role in many parts of mathematics. Of particular importance is the relationship with complex analysis.

By Theorem 17 the fractional linear transformations form the group of all bianalytic transformations of $\mathbf{P}^1(\mathbf{C})$. It is interesting to consider the group of bianalytic transformations that send the unit disc D (which we have also called L in $\mathbf{P}^1(\mathbf{C})$) to itself. It can be calculated explicitly:

**Exercise** Prove that a matrix $\begin{bmatrix} a & b \\ c & d \end{bmatrix}$ satisfies the condition $|z| < 1$ implies $\left|\frac{az+b}{cz+d}\right| < 1$, that is, it sends D to itself if and only if

$$\begin{bmatrix} a & b \\ c & d \end{bmatrix} = \lambda \begin{bmatrix} \alpha & \beta \\ \gamma & \delta \end{bmatrix}$$

with $\alpha = \bar{\delta}$, $\beta = \bar{\gamma}$ and $\lambda = ad - bc$, so $\alpha\bar{\alpha} - \beta\bar{\beta} = 1$.

We will use a different model for D in which it is easier to draw pictures and for which the calculations are slightly easier.

As we saw, the unit disc D can be regarded as the lower hemisphere L of the Riemann sphere. This set L can be rotated round to be a hemisphere that is bounded by the great circle that projects to the real axis. There are two such hemispheres, we will take the one that projects to the upper half plane

$$H = \{z \in \mathbf{C} \mid \operatorname{Im} z > 0\}.$$

This rotation that sends L to H is a fractional linear transformation (by Problem 14, p. 77). In fact it is given by $f\colon D \to H$ where $f(z) = (z+1)/i(z-1)$, its inverse is $f^{-1}(z) = (z-i)/(z+i)$.
Note that $f(0) = i$, $f(1) = \infty$, $f(-1) = 0$, $f(e^{i\theta}) = -\cot(\theta/2)$.

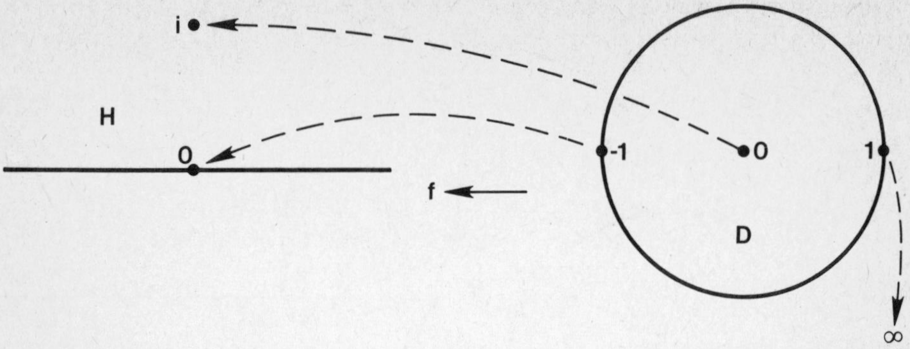


The hyperbolic lines in D become semicircles in H whose centres are on **R** or they become half lines perpendicular to **R** (these are the hyperbolic lines in D meeting the boundary of D at $z = 1$, in H they can be regarded as semicircles with centres at $\infty$). It is straightforward to calculate the group of bianalytic transformations sending H to itself:

**Theorem 21** The fractional linear transformation f defined by $\begin{bmatrix} a & b \\ c & d \end{bmatrix}$ sends H to itself if and only if a, b, c, d are all real multiples of a complex number $\mu$ and $(ad-bc)/\mu^2 > 0$. In other words PSL(2,**R**) is the subgroup of PGL(2,**C**) that sends H to itself.

**Proof** If f maps H to itself, it must map **R** to itself as **R** is the boundary of H, in other words $(az+b)/(cz+d)$ is real for each real z. It is now straightforward to calculate that a, b, c, d must all be multiples of one complex number $\mu$, in particular they can all be chosen to be real, so $f \in$ PGL(2,**R**). If $ad-bc < 0$, H would be mapped to the lower half plane, so $(ad-bc)$ must be positive; dividing by a suitable constant ensures that $ad - bc = 1$ and so $f \in$ PSL(2,**R**).

In this proof we have considered the boundary of H. In $\mathbf{P}^1(\mathbf{C})$ the boundary is a circle and is naturally $\mathbf{P}^1(\mathbf{R}) \subset \mathbf{P}^1(\mathbf{C})$. Every hyperbolic line in H meets $\mathbf{P}^1(\mathbf{R})$in two points. If one of them is at infinity then in the picture it is a straight line perpendicular to **R**.

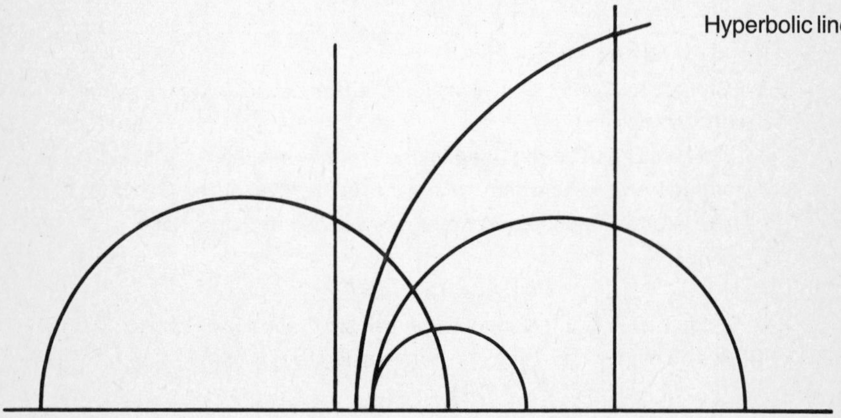

Hyperbolic lines in H

When one regards H in $\mathbf{P}^1(\mathbf{C})$ all hyperbolic lines are the same. The group PSL(2,**R**), being a subgroup of PGL(2,**C**), preserves angles and circles and hence sends hyperbolic lines to hyperbolic lines.

A useful result for handling hyperbolic lines is the following.

**Theorem 22** The four points $z_1, z_2, z_3, z_4 \in \mathbf{P}^1(\mathbf{C})$ lie on a circle if and only if the cross-ratio $(z_1z_2,z_3z_4)$ is real.

**Proof** By Theorem 19, there is a unique $f \in$ PGL(2,**C**) such that $f(z_1) = \infty$, $f(z_2) = 0$, $f(z_3) = 1$. The map f sends circles to circles (Theorem 18) so $f(z_4)$ lies on the circle through $\infty$, 0, 1 but this circle is $\mathbf{P}^1(\mathbf{R})$ so $(z_1z_2,z_3z_4) = f(z_4)$ is real.

Conversely, if $\lambda \in \mathbf{P}^1(\mathbf{R})$ then $\infty$, 0, 1, $\lambda$ lie on a circle, hence so do $z_1, z_2, z_3, z_4$ because $f^{-1}$ preserves circles.

This Theorem could have been proved by a direct calculation.

**Note** We have tacitly assumed in the proof that the four points are distinct. If they are not then they are always concylcic, also the cross-ratio is always 0 or $\infty$ and hence is real.

**Corollary** If $q, r \in \mathbf{P}^1(\mathbf{R})$ then $p \in H$ lies on the hyperbolic line through q, r if and only if $(qr,p\bar{p})$ is real.

**Proof** The hyperbolic line is half of a circle C whose centre is on $\mathbf{P}^1(\mathbf{R})$, and if $p \in C$ so is $\bar{p}$. The converse is also clear.

**Example** We check directly that p lies on the line through 0 and $\infty$ if and only if $(0\infty,p\bar{p})$ is real.

$$(0\infty,p\bar{p}) = (0-p)(\infty-\bar{p})/(0-\bar{p})(\infty-p) = p/\bar{p} = p^2/p\bar{p}$$

so $(0\infty,p\bar{p})$ is real if and only if $p^2$ is real. Let $p = x + iy$, $p^2$ is real if and only if $xy = 0$ but $y \neq 0$ as $p \in H$ so $p^2$ is real if and only if $x = 0$ – as required.

Now we are preparing to prove that the group PSL(2,**R**) is a group of direct isometries of H.

**Theorem 23** The group PSL(2,**R**) is transitive on H, it is even transitive on pairs $(p,\ell)$ such that $p \in \ell$. The stabilizer of a pair $(p,\ell)$ has two elements.

**Proof** Given $p \in \ell$ and $q \in m$, let $\ell$ meet $\mathbf{P}^1(\mathbf{R})$ in a, b and m meet it in c,d.

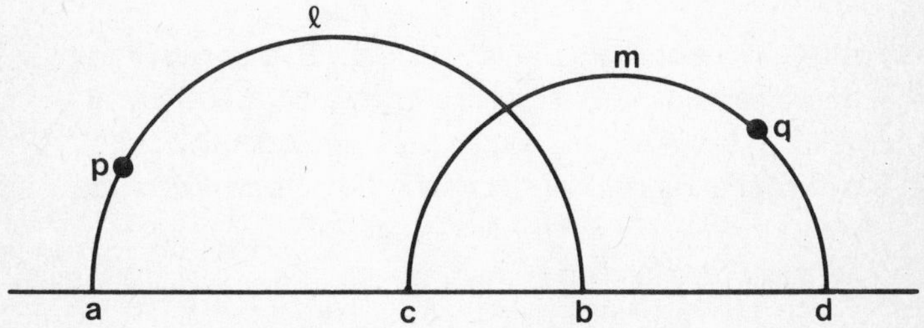


By Theorem 19 there is a unique $f \in$ PGL(2,**C**) such that $fp = q$, $fa = c$, $fb = d$. By Theorem 18, f preserves circles; hence f sends $\ell$ to m. Also f is conformal so it sends a circle orthogonal to $\ell$ to another circle orthogonal to m; hence it sends $\mathbf{P}^1(\mathbf{R})$ to itself; so by the proof of Theorem 21, $f \in$ PSL(2,**R**). This shows that PSL(2,**R**) is transitive on pairs $(p,\ell)$ such that $p \in \ell$.

The only elements of PGL(2,**C**) that send p to p and $\ell$ to $\ell$ are the identity and a transformation g such that gp = p, ga = b, gb = a (again by Theorem 19). Clearly $g^2p = p$, $g^2a = a$ and $g^2b = b$, so $g^2 =$ identity.

Our conclusion is that PSL(2,**R**) is analogous to a group of direct isometries of H. To get the (larger) group of all isometries one adds the isomorphism $z \to 1/\bar{z}$ and considers the group of transformations generated by PSL(2,**R**) and this extra element. Notice that $z \to 1/\bar{z}$ also preserves circles and angles (up to sign). It sends $\mathbf{P}^1(\mathbf{R})$ to itself and i to itself. When we need to use this larger group we denote it by PSL*(2,**R**).

We could now proceed to define a metric on H by using our previous formula. If p, q ∈ H and the hyperbolic line through p, q meets $\mathbf{P}^1(\mathbf{R})$ in a, b then define $d(p,q) = |\log(pq,ab)|/2$. This distance is preserved by the whole group PSL*(2,**R**) and it can be verified that this is the group of all isometries of the Poincaré model H. For ease of calculation and for novelty we take a different approach.

**The Local Metric**

Given a point p ∈ H, what is the element of distance $ds_p$ at p? Such an element of distance is called a local metric and is exactly what is required to define geodesics locally. Clearly $ds_p$ must be given by an expression of the following form

$$ds_p{}^2 = a(p)dx^2 + 2b(p)dxdy + c(p)dy^2,$$

for some real valued functions a, b, c defined on H. It is somewhat more convenient to rewrite this as

$$ds_z{}^2 = A(z)dz^2 + B(z)dzd\bar{z} + C(z)d\bar{z}^2 \text{ for } z \in H \text{———————} (*)$$

**Exercise** By comparing these two expressions; prove that B(z) is real and that $A(z) = \overline{C(z)}$. Use these facts to simplify the following proof.

**Theorem 24** The only metrics of the form (*) that are invariant under the group PSL(2,**R**) are of the form $ds_z{}^2 = k(dx^2+dy^2)/y^2$ where $z = x + iy$ and k is a positive constant.

**Proof** The metric must be invariant under the transformation $z \to z + \lambda = w$ for $\lambda \in \mathbf{R}$. In this case dz = dw and it is easy to see that then the functions A, B, C of (*) are independent of x.

The metric is also invariant under $z \to \lambda z = w$ for $\lambda > 0$. In this case $dw = \lambda dz$ so one gets that

$$A(\lambda z)\lambda^2 dz^2 = A(z)dz^2$$

and hence A is homogeneous of degree –2 in z. Similarly B, C are also homogeneous of degree –2 in z, but all three are independent of x, so there are constants A, B, C so that

$$A(z) = A/y^2, \qquad B(z) = B/y^2, \qquad C(z) = C/y^2.$$

Finally, the metric is invariant under $z \to -1/z = w$. In this case $dw = dz/z^2$ and

$$w = u + iv = -(x-iy)/|z|^2,$$

so $y = |z|^2 v$. We have $A(w) = A/v^2$ and $A(z) = A/y^2$, so

$$Adz^2/y^2 = Adz^2/v^2z^4 = A|z|^4dz^2/y^2z^4.$$

Hence $(z^4-|z|^4)A = 0$ for all z ∈ H, so A = 0. Similarly C = 0. So $ds_z{}^2 = B(dx^2+dy^2)/y^2$. Clearly $B > 0$.

When one has a metric on a locally Euclidean space (more precisely, on a smooth manifold) one can define a local metric ds by considering the lengths of small line segments near a point. This can be viewed as differentiation. Conversely if a local metric ds is given on such a space, one can define a global metric by integration. The distance between two points p, q is defined to be

$$\inf_{\alpha} \int_{\alpha} ds,$$

where the infimum is taken over all paths α from p to q. This is often called the geodesic distance

– the length of the shortest path between two points. This all works well when the space is compact but when the space is not compact the distance may not be realised as the length of any curve. For example, consider $\mathbf{R}^2 \setminus \{0\}$ and the points to be (1,0) and (–1,0).

### Areas

Similarly one can consider area elements, in our case the area element at the point $z = x + iy \in H$ is clearly $dxdy/y^2$. We will now use this formula to calculate the area of a hyperbolic polygon in H, that is, an area bounded by a finite number of hyperbolic lines of H. An example of such a polygon is the following:

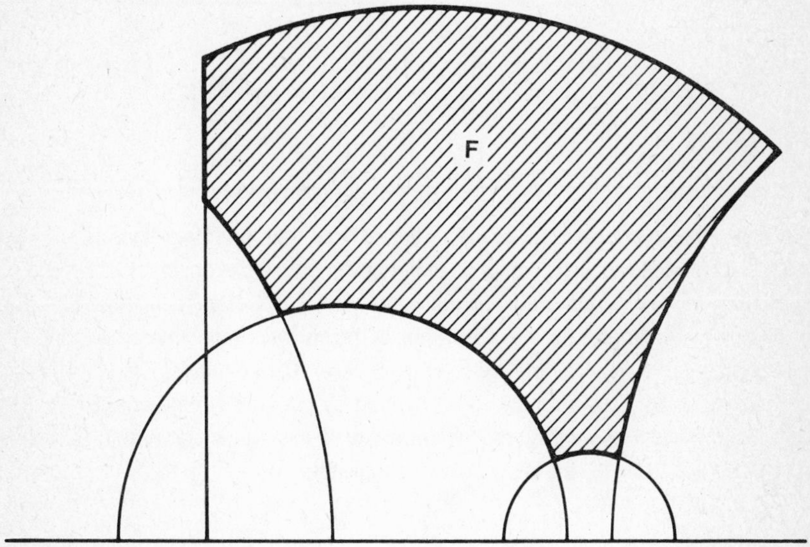


By our definition, the area is

$$\int_F \frac{dxdy}{y^2}$$

To evaluate this we use Green's theorem

$$\int_F \left(\frac{\partial v}{\partial x} - \frac{\partial u}{\partial y}\right) dxdy = \int_{\partial F} udx + vdy$$

by letting $u = 1/y$ and $v = 0$, so $\partial u/\partial y = -1/y^2$. This gives

$$\int_F \frac{dxdy}{y^2} = \int_{\partial F} \frac{dx}{y}$$

To evaluate this integral one calculates the value of the right hand side as one traverses a segment of a hyperbolic line – which is an edge of F. Let L be a segment of a hyperbolic line (whose euclidean radius is r: L is the portion between angles $\beta$ and $\alpha$ as shown below)

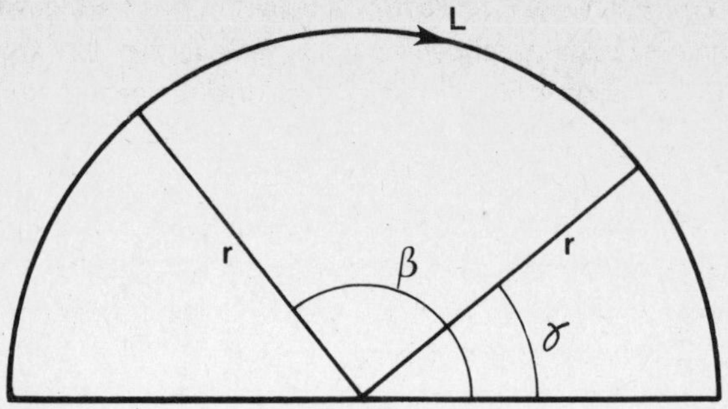


Then one has that

$$\int_L \frac{dx}{y} = \int_\beta^\gamma \frac{d(r\cos\theta)}{r\sin\theta} = -\int_\beta^\gamma d\theta = \beta - \gamma;$$

hence this integral is independent of the radius r. Note that the integral vanishes if L is a vertical line. This can be seen from this calculation or directly because dx vanishes.

With this preliminary we can now calculate the area of F. Suppose its edges are $L_1, L_2, \ldots, L_m$ with $L_k$ defined by the angles $\beta_k$, $\gamma_k$ – the radius of the corresponding hyperbolic line is not needed. If $L_k$ is a segment of a vertical line we take $\beta_k = \gamma_k$ $(= 0)$. The conclusion is that the hyperbolic area of F is $\sum_{k=1}^{n} (\beta_k - \gamma_k)$. This sum can be simplified by considering the change in the direction of the outward normal to F as we traverse the boundary $\partial F$ in an anticlockwise direction. Let the vertex at which $L_k$ meets $L_{k+1}$ be $A_k$ and let the internal angle there be $\alpha_k$.

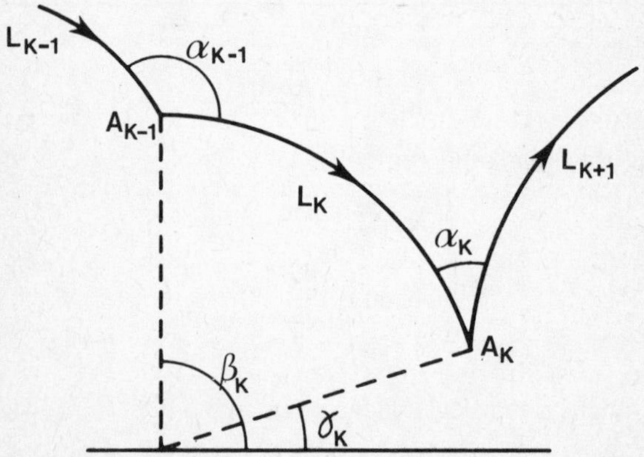


As we traverse $L_k$ the change in the direction of the outward normal is $\gamma_k - \beta_k$. At the vertex $A_k$ the change is $\pi - \alpha_k$. So the total change as we traverse around F is

$$m\pi + \sum_{k=1}^{m} (\gamma_k - \beta_k - \alpha_k)$$

But the total change is clearly $2\pi$, so one has the equation

$$2\pi = m\pi + \sum_{k=1}^{m} (\gamma_k - \beta_k) - \sum_{k=1}^{m} \alpha_k$$

But

$$\text{Area}(F) = \sum_{k=1}^{m} (\beta_k - \gamma_k)$$

so one deduces that

$$\text{Area}(F) = \pi(m-2) - \sum_{k=1}^{m} \alpha_k$$

So the area of a hyperbolic polygon depends only on its internal angles.

In particular a hyperbolic triangle $\triangle$ whose angles are $\alpha$, $\beta$, $\gamma$ has its area given by

$$\text{Area}(\triangle) = \pi - (\alpha+\beta+\gamma).$$

It is interesting to compare this with the area of an elliptic triangle (given on page 64) which is $(\alpha+\beta+\gamma)-\pi$.

This gives a way in which the three plane geometries can be distinguished and is closely connected with the 'curvature' of the space:

Given any triangle of a plane geometry, let its angles be $\alpha$, $\beta$, $\gamma$. One can decide which plane geometry it belongs to by considering the sign of $(\alpha+\beta+\gamma)-\pi$. Similar measurements can be carried out in three dimensional space – this has been done to try to discover the curvature of the space that we live in. Non zero results that are outside experimental error have not been found – so that the curvature may be zero. The study of invariants such as curvature is called differential geometry and plays an important role in general relativity.

We have the summarising table:

| $(\alpha+\beta+\gamma)-\pi$ | **Geometry** | **Curvature** |
|---|---|---|
| $<0$ | Hyperbolic | negative |
| $>0$ | Elliptic | positive |
| $0$ | Euclidean | flat |

Unfortunately, it is cumbersome to calculate the trigonometry of hyperbolic figures using the Poincaré model. It is easier to use a slight variant of the Beltrami model, this we now do.

**Trigonometry**

We are going to study hyperbolic and elliptic (or spherical) trigonometry (I presume that the reader knows about euclidean trigonometry but it will not be used here). Some preliminary work is common to both cases and anyway the two are very similar.

From now on we will have a fixed non-degenerate bilinear form on $\mathbf{R}^3$ which will be denoted x.y (rather than the more cumbersome $x^tAy$ used in the discussion of conics). Using the diagonalization of quadratic and bilinear forms over $\mathbf{R}$, choose a basis $e_1, e_2, e_3$ for $\mathbf{R}^3$ so that

$$e_i.e_j = \varepsilon_i\delta_{ij} \qquad \text{with} \qquad \varepsilon_i = \pm 1.$$

Three points $x_1, x_2, x_3 \in \mathbf{R}^3$ determine a $3 \times 3$ matrix $(\xi_{ij})$ by $x_i = \Sigma\, \xi_{ij}e_j$. Define

$$\text{vol}(x_1,x_2,x_3) \qquad \text{to be} \qquad \det(\xi_{ij}).$$

**Note** $\text{vol}(x_1,x_2,x_3)^2$ is independent of the choice of basis because it equals $\det(x_i{}^t x_j)$. (In the case of the standard inner product, $\text{vol}(x_1x_2x_3)$ is the volume of the parallelipiped with edges $0x_1$, $0x_2$, $0x_3$.) To define the vector product we use the following general result:

**Lemma** Suppose that x.y is a non-degenerate bilinear form on $\mathbf{R}^n$, then for every linear mapping $f: \mathbf{R}^n \to \mathbf{R}$ there is a unique $x_f \in \mathbf{R}^n$ such that $f(y) = x_f.y$ for all $y \in \mathbf{R}^n$.

**Proof** Choose a basis $e_1, e_2, \ldots, e_n$ for $\mathbf{R}^n$ such that

$$e_i.e_j = \varepsilon_i\delta_{ij} \qquad \text{for all} \qquad i, j.$$

Define $x_f$ to be $\sum_{i=1}^{n} \varepsilon_i\lambda_i e_i$ where $\lambda_i = f(e_i)$ for each i.

Then $x_f.e_j = \sum_{i=1}^{n} \varepsilon_i\lambda_i e_i.e_j = \sum_{i=1}^{n} \varepsilon_i\lambda_i\varepsilon_i\delta_{ij} = \lambda_j = f(e_j)$.

By linearity $x_f.y = f(y)$ for all y.

The uniqueness of $x_f$ follows from the non-degeneracy of x.y.

**Corollary** For every $x,y \in \mathbf{R}^3$, there is a unique vector $x \times y \in \mathbf{R}^3$ such that $(x\times y).z = \mathrm{vol}(x,y,z)$ for all $z \in \mathbf{R}^3$.

**Proof** The map $z \to \mathrm{vol}(x,y,z)$ is linear and so the result follows.

**Proposition** i) $(\lambda x) \times y = \lambda(x\times y)$

ii) $y \times x = -x \times y$

iii) $(x\times y).z = (y\times z).x = (z\times x).y = x.(y\times z) = y.(z\times x) = z.(x\times y)$.

These all follow from standard properties of the determinant.

Now define $\varepsilon$ to be $\varepsilon_1\varepsilon_2\varepsilon_3$ and note that $\varepsilon$ is independent of the choice of the basis $e_1$, $e_2$, $e_3$ because it is the sign of the determinant of the matrix of the bilinear form.

**Proposition** $x \times (y\times z) = \varepsilon[(x.z)y - (x.y)z]$.

This can be proved by following a proof of the usual case. The proof that follows is just one of many such proofs.

**Proof** It is clear that both sides are linear in each of the three variables. It is therefore enough to check that the formula holds when the three variables are taken to be the basis vectors. By the symmetry of $e_1$, $e_2$, $e_3$ we may take x to be $e_1$. If either a) $y = z$ or b) neither y nor z equals $e_1$ then the left hand side vanishes; it is also clear that the right hand side vanishes in both these cases. When y and z are interchanged, both sides change sign. Hence it suffices to consider $y = e_1$ and $z = e_2, e_3$. The formula is easily checked in both these cases.

**Corollary** $(x\times y).(y\times z) = \varepsilon[(x.y)(y.z) - (x.z)(y.y)]$.

**Proof** $(x\times y).(y\times z) = [y \times (y\times z)].x = \varepsilon[(y.z)(y.x) - (y.y)(z.x)]$

In both the applications given in these notes $\varepsilon$ will be $+1$ so the reader need not worry unduly about it.

## Hyperbolic Trigonometry

We will work with the Berltrami model whose absolute conic is

$$x_1^2 - x_2^2 - x_3^2 = 0.$$

The inside of the conic is the set of points $[x_1:x_2:x_3]$ in $\mathbf{P}^2(\mathbf{R})$ for which $x_1^2 - x_2^2 - x_3^2 > 0$ and this is the model $H \subset \mathbf{P}^2(\mathbf{R})$ for the hyperbolic plane, the lines are the intersections of the projective lines with H.

Representatives for the points of H can be chosen on the hyperboloid $x_1^2 - x_2^2 - x_3^2 = 1$ in $\mathbf{R}^3$, this hyperboloid has two sheets, each point of H has two such representatives. We will always consider the representative on the sheet with $x_1 \geqslant 1$ (the other sheet has $x_1 \leqslant -1$). The points on the outside of the absolute conic each have two representatives on the hyperboloid of one sheet

$$x_1^2 - x_2^2 - x_3^2 = -1.$$

There is no simple way of singling out one of these representatives.

sinh c sinβ and the result follows.
is a line in H and b is a point not on $a^\perp$, then there is a perpendicular from b to $a^\perp$
e from b to $a^\perp$ is d where sinh d = a.b.

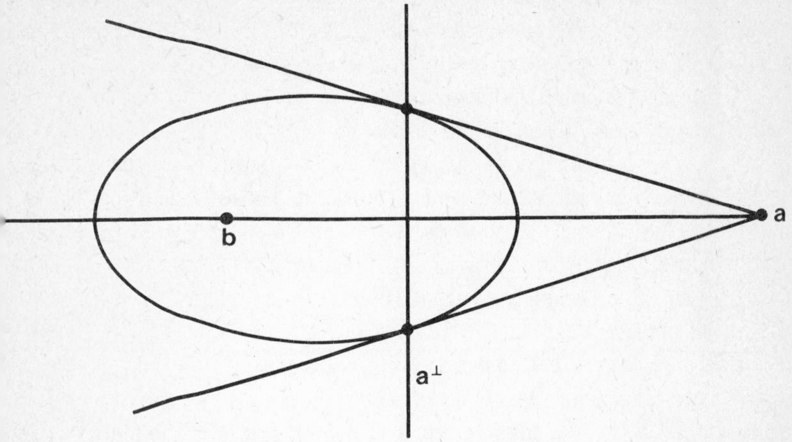


perpendicular to $a^\perp$, because (b×a).a vanishes.
point where $a^\perp$ meets the line ab. Then $p = \lambda b + \mu a$ with p.a = 0, p.p = 1 and
xpanding everything out yields

$$\lambda(\lambda+\mu(a.b)) = 1 \qquad \lambda(a.b)-\mu = 0 \qquad \lambda + \mu(a.b) = \cosh d.$$

d and (a.b) = sinh d.
also gives us a very famous formula of hyperbolic geometry.
as above, then there are two lines through b that are parallel to $a^\perp$, they meet $a^\perp$
conic.

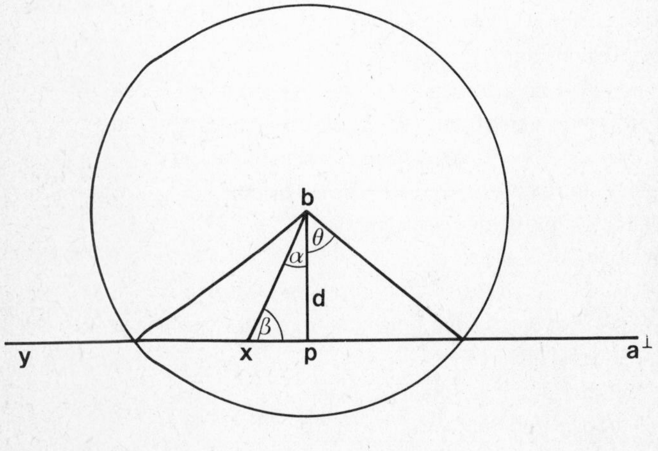


s make equal angles θ with the perpendicular bp from b to $a^\perp$. This angle is called
**rallelism**, it is given by the following equivalent formulae:



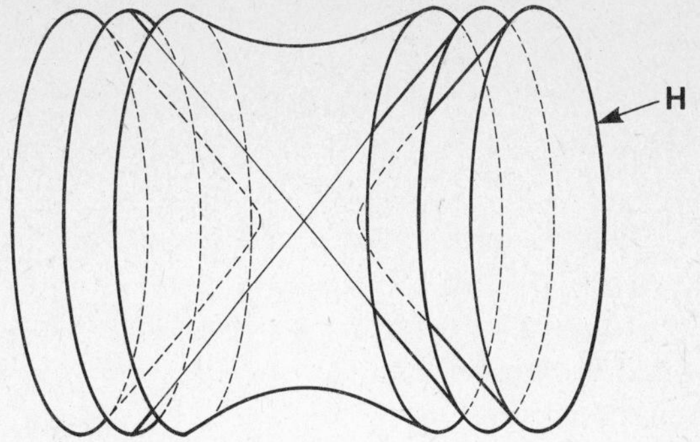


We will always have in mind the fixed bilinear form

$$x.y = x_1y_1 - x_2y_2 - x_3y_3.$$

Two distinct points x, y ∈ $\mathbf{P}^2(\mathbf{R})$ determine a plane {x,y} in $\mathbf{R}^3$, {x,y} is spanned by the representatives of x and y. The bilinear form restricted to the plane {x,y} can be of one of three types:

1. Negative definite (type (0,2), that is, no positive terms and two negative terms in its diagonalized form). In this case the line xy does not meet the absolute conic.

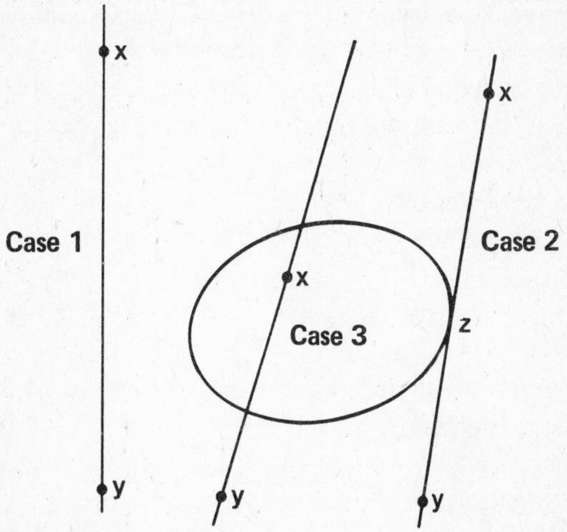


2. Negative semidefinite (type (0,1)). In this case the line xy is tangent to the absolute conic (at z, say). The representatives of z are the non-zero vectors in the plane {x,y} that satisfy $\tilde{z}.\tilde{z} = 0$.
3. Indefinite (type (1,1)). In this case the line xy crosses the absolute conic.

Using the discriminant of a quadratic form q on $\mathbf{R}^2$, it is easy to prove the following algebraic criteria

| **type of q** | **criteria** |
|---|---|
| (2,0) | $x.x > 0$ and $(x.x)(y.y) > (x.y)^2$ |
| (1,1) | $(x.x)[(x.y)^2-(x.x)(y.y)] > 0$ |
| (0,2) | $(x.x) < 0$ and $(x.y)(y.y) > (x.y)^2$ |
| degenerate | $(x.x)(y.y) = (x.y)^2$. |

**Lines and Polarity**

**Lemma** If $\ell$ is a line in H, its pole a lies outside C.

**Proof** Suppose $x \in \ell$ and $x.x > 0$, then there is an $\alpha$ such that $x + \alpha a \in C$ so

$$0 = (x+\alpha a).(x+\alpha a) = x.x + 2\alpha a.x + \alpha^2 a.a; \quad a.x = 0 \text{ and } x.x > 0$$

hence $a.a < 0$.

Conversely, if a is outside C, its polar line $\ell$ must cross C: otherwise every $x \in \ell$ satisfies $x.x < 0$. Every point in $\mathbf{P}^2(\mathbf{R})$ can be written as $\alpha a + \beta x$ for some $x \in \ell$ and then $(\alpha a+\beta x).(\alpha a+\beta x) < 0$ because $a.x = 0$, hence the form would have to be negative definite on the whole of $\mathbf{R}^3$, which it is not. This argument also proves

**Proposition** Consider the point $x \in \mathbf{P}^2(\mathbf{R})$, then

x lies inside C if and only if its polar does not meet C

x lies on C if and only if its polar is tangent to C

x lies outside C if and only if its polar crosses C.

As in the Euclidean case, the space of all lines in H is a Möbius band because there is a homeomorphism (obtained by taking the pole) between the space of lines and the space of points lying outside C; the outside of a conic is homeomorphic to a Möbius band.

**Notation** A line in H will be denoted by $a^\perp$ where a is its pole.

The two choices of the representatives of the pole of the line $\ell$ on the hyperboloid $x.x = -1$ correspond to the two directions of $\ell$. If we wish to consider the direction of a line $\ell$ then we will write it as $\tilde{a}^\perp$ for a particular choice $\tilde{a}$ for the representative of the pole a of $\ell$.

**Proposition** The lines $a^\perp$, $b^\perp$ meet, are parallel or utlraparallel according as the plane $\{a,b\}$ is of type (0,2), (0,1) or (1,1).

**Proof** The plane $\{a,b\}$ represents the polar of $a^\perp \cap b^\perp$. The lines $a^\perp$, $b^\perp$ meet if and only if $a^\perp \cap b^\perp$ is inside C, they are parallel if and only if $a^\perp \cap b^\perp$ lies on C and they are ultraparallel if and only if $a^\perp \cap b^\perp$ lies outside C.

If $a^\perp$, $b^\perp$ meet then $\{a,b\}$ is of type (0,2) and $a.a = b.b = -1$, one has that $(a.b)^2 < 1$. Hence the following is well defined:

**Definition** If $a^\perp$, $b^\perp$ meet then the **angle** between them is defined to be $\theta$ where $\cos\theta = a.b$.

We now define the distance between two points of H:

**Proposition** If $x, y \in H$ then $x.y \geqslant 1$.

(Note that H is taken ambiguously to be the inside of the absolute conic or $\{x \in \mathbf{R}^3 | x.x = 1 \text{ and } x_1 \geqslant 1\}$.)

**Proof** Let $x, y \in H$ and $m = (x+y)/2$ be the mid point of the line segment xy in $\mathbf{R}^3$. By the convexity of the hyperboloid, $m.m \geqslant 1$. Hence $(x+y).(x+y) \geqslant 4$, so $x.y \geqslant 1$.

Now we can **define the metric** d on H.

**Definition** If $x.y \in H$, $\cosh d(x,y) = x.y$.

**Proposition** If xyz is a hyperbolic triangle then

$$\cosh b = \cosh a \cosh c - \sinh a \sinh c \cos\beta \qquad (1)_b$$

$$\cos\beta = -\cos\alpha \cos\gamma + \cosh b \sin\alpha \sin a \qquad (1)_\beta .$$

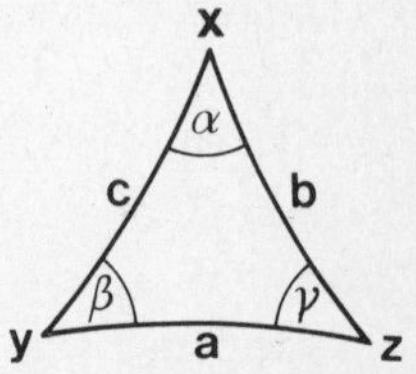


There are identical formulae $(1)_a$, $(1)_c$, $(1)_\alpha$, $(1)_\gamma$ for cosh a, cosh c, cos$\alpha$ and cos$\gamma$ respectively.

In the proof we will need two preliminary results.

**Lemma** Let $x, y \in H$ be distance d apart, so $x.y = \cosh d$, t
it is the pole of the directed line xy.

**Proof** Clearly $(x \times y).x = (x \times y).y = 0$, so the line xy is the pola

Also $(x\times y).(x\times y) = -(x\times y).(y\times x) = 1 - (x.y)^2$ by the Corolla

Hence $(x\times y).(x\times y) = -\sinh^2 d$ as required.

**Convention** If x, y, z are three non-collinear points in H th
the line segments xy and yz has the same sign as $(x\times y).(y\times z)$
So $\cos\theta = (x\times y).(y\times z)/\sinh d(x,y) \sinh d(x,z)$.

**Proof of the proposition** Consider the identity

$$(x\times y).(y\times z) = (x.y)(y.z) - (x.z$$

given by the Corollary on page 92 and interpret the terms. The

$$\cos\beta \sinh c \sinh a$$

and the right hand side is cosh c cosh a – cosh b. The formula (

**Corollary** The triangle inequality is satisfied by d.

**Proof** By the proposition

$$\cosh b \leqslant \cosh a \cosh c + \sinh a \sinh c$$

But cosh x is an increasing function for positive x, hence $b \leqslant a$

To prove formula $(1)_\beta$, we consider the poles of the side
xy is $(x\times y)/\sinh c$. Let $(x\times y) = \zeta \sinh c$, $(z\times x) = \eta \sinh b$ and (y

$$(\xi\times\eta).(\eta\times\zeta) = (\xi.\eta)(\eta.\zeta) - (\xi.$$

But $(\xi\times\eta).(\xi\times\eta) = -(\xi.\eta)^2 + 1 = \sin^2\gamma$

So $z \sin\gamma . x \sin\alpha = \cos\gamma \cos\alpha + \cos\beta$, and as $x.z = \cosh b$, for

**Proposition** If xyz is a hyperbolic triangle right angled at y,

i) $\cosh b = \cosh a \cosh c$,

ii) $\cos\alpha = \cosh a \sin\gamma$,

$\cos\gamma = \cosh c \sin\alpha$,

iii) $\sinh a = \sinh b \sin\alpha$,

$\sinh c = \sinh b \sin\gamma$.

**Proof** The formulae i) and ii) follow immediately from the ab

To prove formula iii), consider $(1)_a$

$$\cosh a = \cosh b \cosh c - \sinh b \, s$$

substitute cosh a cosh c for cosh b from i), use $\cosh^2 c = 1 +$

$$\cosh a \sinh c = \sinh b \, co$$

Now substitute cosh a sin$\gamma$ for cos$\alpha$ from ii) and divide by cos

**Proposition** (The analogue of the law of sines).

If xyz is a hyperbolic triangle, then

$$\sinh a/\sin\alpha = \sinh b/\sin\beta = si$$

To prove this we need to drop a perpendicular from x to
be done after giving the rest of the proof.

**Proof** Let p be the foot of the perpendicular and $h = d(x,p)$. By iii) above applied to the two right angled triangles, one gets

$$\sinh h = \sinh c \sin\beta$$

and $$\sinh h = \sinh b \sin\gamma$$

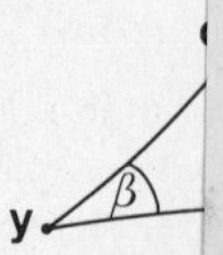




So sinh b

**Lemma**

and the di

**Proof**

The line

Let p be

cosh d = b.

So λ = 1/co

This lem

Let a, b

on the absc

These two li

the **angle** of

$$\sin\theta = \operatorname{sech} d, \qquad \tan\theta/2 = e^{-d}.$$

**Proof** Consider the right angled triangle bpx. By the formulae for such a triangle,

$$\cos\beta = \cosh d \sin\alpha.$$

In the limit, as x tends to y, the angle $\beta$ tends to zero, hence

$$1 = \cosh d \sin\theta.$$

This gives the first formula and the fact that the two angles of parallelism are equal.

Let $t = \tan\theta/2$, then $\sin\theta = 2t/(1+t^2)$.

But $\operatorname{sech} d = 2/(e^d+e^{-d})$.

Hence $(e^d+e^{-d}) = 1 + t^2$ so $t = e^d$ or $e^{-d}$.

But $0 < \theta < \pi/2$ so $0 < \tan\theta/2 < 1$, hence $t = e^{-d}$.

**Isometries**

**Definition** The group $O(3')$ consists of the linear transformations of $\mathbf{R}^3$ that preserve the form x.y, that is,

$$O(3') = \{T \in GL(3,\mathbf{R}) \mid Tx.Ty = x.y \text{ for all } x,y \in \mathbf{R}^3\}.$$

The elements of $O(3')$ define mappings $H \to H$ that are isometries of H with the metric defined by $\cosh d(x,y) = x.y$. The elements of $O(3')$ also preserve angles.

**Proposition** The group $O(3')$ acts transitively on pairs $(x,\ell)$ where x is a point of the line $\ell$.

**Proof** If x is any point in H and $y \in x^\perp$, let $z = y^\perp \cap x^\perp$. Consider the linear transformation defined by $Tx = (1,0,0)$, $Ty = (0,1,0)$ and $Tz = (0,0,1)$. As $y, z \in x^\perp$ and $x^\perp$ does not meet the absolute conic we have $y.y = z.z = -1$. Hence $T \in O(3')$. The line $T\ell$ passes through $(1,0,0)$. It is clear that $O(3')$ contains all rotations about the $x_1$ axis, so there is an $S \in O(3')$ such that $S(1,0,0) = (1,0,0)$ and $ST\ell$ is any given line through $(1,0,0)$.

The image $PO(3')$ of $O(3')$ in $PGL(3,\mathbf{R})$ preserves the conic and preserves cross-ratios. Hence $PO(3')$ is the group of isometries of H.

In hyperbolic geometry there are many other interesting constructions and we end by considering some of them.

**Proposition** For every $n \geq 3$, there is a regular n-gon with angle $\alpha$ for any $\alpha$ satisfying

$$0 < \alpha < (n-2)\pi/n.$$

**Proof** Consider the Poincaré disc model – in which the lines are arcs of circles perpendicular to the edge of the disc. Take n rays emanating from the centre, with equal angles between them. Take points $P_1, P_2, \ldots, P_n$ on each ray at equal distances d from the centre. Then $P_1 P_2 \ldots P_n$ is a regular n-gon. When d is small, the angle $\alpha$ is close to $(n-2)\pi/n$ and when d is large, the angle $\alpha$ becomes close to 0. By continuity any value of $\alpha$ between 0 and $(n-2)\pi/n$ is attained for some value of d.

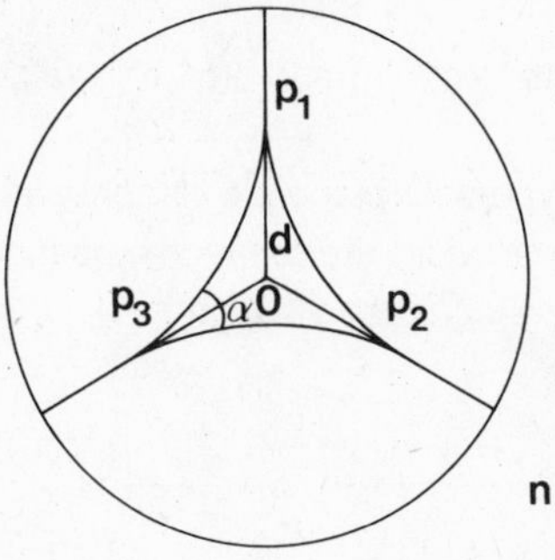


n = 3

**Corollary** For every $n \geqslant 5$, there is a regular n-gon all of whose angles are right angles.

Of course, these particular n-gons we have constructed can be moved around using elements of $O(3')$ to give a large number of others.

As a further example we consider the trigonometry of a right angled hexagon.

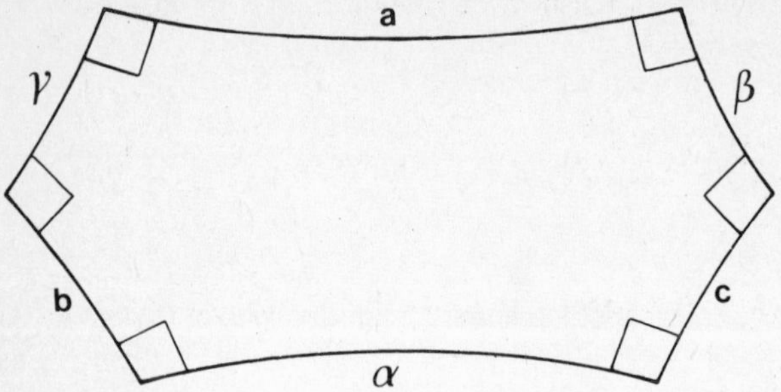


**Proposition** In a right angled hexagon with side lengths as shown one has

$$\cosh \beta = -\cosh \alpha \cosh \gamma + \sinh \alpha \sinh \gamma \cosh b.$$

**Proof** As the angles are right angles, we have that $a^{\perp}.\beta^{\perp} = a^{\perp}.\gamma^{\perp} = 0$ where $a^{\perp}$ denotes the pole of the side labelled a, etc. So $\beta^{\perp}, \gamma^{\perp}$ both lie on the line a. So we have:

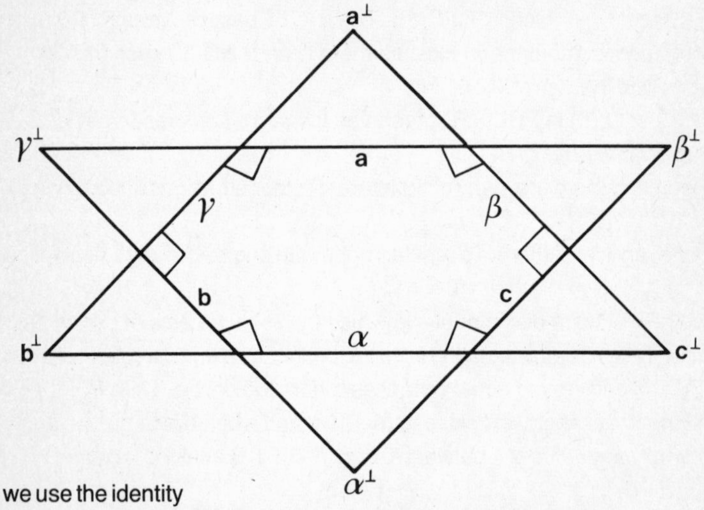


As usual we use the identity

$$(a^{\perp} \times b^{\perp}).(b^{\perp} \times c^{\perp}) = (a^{\perp}.b^{\perp})(b^{\perp}.c^{\perp}) - (a^{\perp}.c^{\perp})(b^{\perp}.b^{\perp})$$

but we need a preliminary result

**Lemma** Two lines $\ell$, m in the hyperbolic plane are ultraparallel if and only if they have a common perpendicular. Their distance d apart along the common perpendicular is given by

$$\cosh d = \ell^{\perp}.m^{\perp} .$$

**Proof** Remember $\ell$, m are ultraparallel if and only if they meet outside C. Let $p = \ell \cap m$, then $p.\ell^{\perp} = 0$ so $p^{\perp}$ is orthogonal to $\ell$, similarly it is orthogonal to m hence $p^{\perp}$ is a common perpendicular.

Conversely suppose n is a common perpendicular then $n^{\perp}.\ell^{\perp} = 0$ and $n^{\perp}.m^{\perp} = 0$ so $n^{\perp} = \ell \cap m$, but $n^{\perp}$ is outside C, so $\ell$, m are ultraparallel.

Now let x, y be the points $\ell \cap n$ and $m \cap n$. We must show that $x.y = \ell^{\perp}.m^{\perp}$.

Clearly $\ell^{\perp} = \lambda[(x.y)x-(x.x)y]$ and $m^{\perp} = \mu[(y.y)x-(x.y)y]$.

The constants $\lambda$, $\mu$ are determined by $\ell^{\perp}.\ell^{\perp} = m^{\perp}.m^{\perp} = -1$ and are $\lambda^2 = \mu^2 = -1/(1-(x.y)^2)$

But $\ell^{\perp}.m^{\perp} = \lambda\mu[1-(x.y)^2](x.y)$. The result follows.

So $a^{\perp}.b^{\perp} = \cosh\gamma$, $a^{\perp}.c^{\perp} = \cosh\beta$, $b^{\perp}.c^{\perp} = \cosh\alpha$ and $b^{\perp}.a^{\perp} = -1$.

$$(a^{\perp}\times b^{\perp}).(a^{\perp}\times b^{\perp}) = -(a^{\perp}.b^{\perp})(b^{\perp}.a^{\perp}) + (b^{\perp}.a^{\perp})(b^{\perp}.a^{\perp})(b^{\perp}.b^{\perp})$$
$$= 1 - \cosh^2\gamma = -\sinh^2\gamma$$

and $(b^{\perp}\times c^{\perp}).(b^{\perp}\times c^{\perp}) = -\sinh^2\alpha$.

Therefore $(a^{\perp}\times b^{\perp}).(b^{\perp}\times c^{\perp}) = \sinh\alpha \sinh\gamma \cosh b$ and the formula follows.

Spherical and elliptic trigonometry can be treated in a similar manner, for completeness we now discuss this.

## Elliptic Trigonometry

The elliptic plane is denoted by $P^2$ and representatives of its points are vectors $x \in S^2 = \{x \in \mathbf{R}^3 | \, \|x\| = 1\}$. Both x and $-x$ represent the same point of $P^2$. The distance between two points x, $y \in P^2$ is the (acute) angle between the lines Ox, Oy in $\mathbf{R}^3$. In symbols, $\cos d(x,y) = |x.y|$ where x.y is the (usual) Euclidean inner product of $\mathbf{R}^3$.

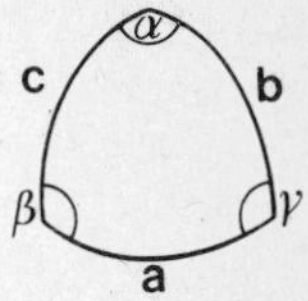


We now study the trigonometry of an elliptic triangle xyz.

Let $a = d(y,z)$, etc. and $\alpha$ = the angle at x. etc. as shown.

**Theorem** In an elliptic triangle $\cos b = \cos a \cos c + \sin a \sin c \cos\beta$ -①$_b$.

There are similar formulae ①$_a$ and ①$_c$ for cos a and cos c.

**Proof** We use the identity $(x\times y).(y\times z) = (x.y)(y.z) - (x.z)(y.y) = \cos c \cos a - \cos b$

Now $\|x\times y\|^2 = -(x\times y).(y\times x) = -(x.y)^2 + (x.x)(y.y) = -\cos^2 c + 1 = \sin^2 c$

and $\|y\times z\|^2 = \sin^2 a$.

The angle between $(y\times x)$ and $(y\times z)$ is $\beta$.

So $(x\times y).(y\times z) = -\sin a \sin c \cos\beta$, giving the result.

**Corollary** The triangle inequality: $d(x,z) \leq d(x,y) + d(y,z)$.

**Proof** $\cos b = \cos a \cos c + \sin a \sin c \cos\beta \geq \cos a \cos c - \sin a \sin c = \cos(a+c)$.

Hence $b \leq a + c$ because cos is a decreasing function of x for $0 \leq x \leq \pi$.

One can obtain further formulae by considering the polar triangle. (Remember that if the line is the equator of the sphere the pole is the north (or south) pole, this probably explains the use of the words pole and polar.) The edges of the polar triangle of xyz are the polars of x, y, z. Let the polar triangle be x′y′z′ where x′ is the point where the polars of y and z meet etc. Let its edge lengths be a′, b′, c′ and its angles be $\alpha'$, $\beta'$, $\gamma'$.

**Proposition** $\alpha + a' = \pi$, $\beta + b' = \pi$ and $\gamma + c' = \pi$.

**Proof** Suppose the lines xy and xz meet y′z′ in p, q respectively. Then $\alpha$ equals d(p,q).

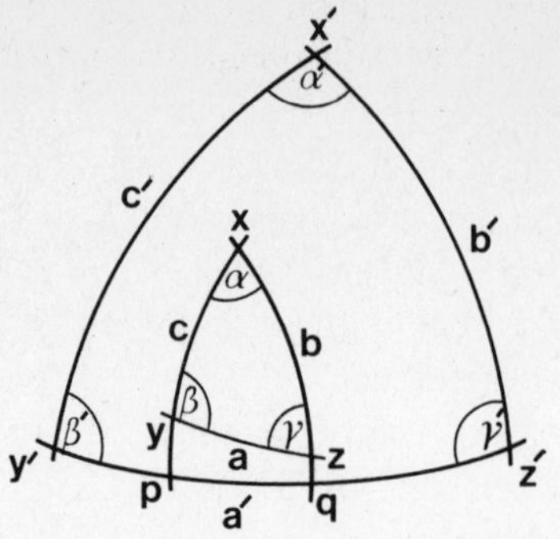


The polars of x and z meet at $y'$, the point q is collinear with x and z so its polar passes through $y'$, hence the $d(q,y')$ is $\pi/2$. Similarly $d(p,z') = \pi/2$ so $d(p,q) + d(y',z') = \pi$, proving the result.

**Corollary** $\cos\beta = -\cos\alpha\cos\gamma + \sin\alpha\sin\gamma\cos b$ ———————— $①_\beta$
and also $①_\alpha$, $①_\gamma$.

**Proof** Apply the trigonometric identity that we have already proved to the polar triangle and use the proposition.

When we apply these formulae to a right angled triangle we get:

**Proposition** If xyz is a triangle right angled at y, then

i) $\cos b = \cos a \cos c$

ii) $\cos\alpha = \cos a \sin\gamma$
$\cos\gamma = \cos c \sin\alpha$

iii) $\sin a = \sin b \sin\alpha$
$\sin c = \sin b \sin\gamma$.

**Proof** i) Put $\cos\beta = 0$ in $①_b$.
ii) Put $\cos\beta = 0$, $\sin\beta = 1$ in $①_\alpha$.
iii) Use $①_a$: $\cos a = \cos b \cos c + \sin b \sin c \cos\alpha$

Substitute cos a cos c for cos b (from i)) and use $\cos^2 c = 1 - \sin^2 c$. Divide by sin c to get $\cos a \sin c = \sin b \cos\alpha$. Substitute $\cos a \sin\gamma$ for $\cos\alpha$ (from ii)) and divide by cos a to give the result.

**Proposition** (Law of sines)

Let xyz be any triangle, then

$$\sin a/\sin\alpha = \sin b/\sin\beta = \sin c/\sin\gamma.$$

**Proof** Drop the perpendicular from x to yz

By iii) of the previous proposition

$$\sin h = \sin c \sin\beta$$

and also $\sin h = \sin b \sin\gamma$.

So $\sin b \sin\gamma = \sin c \sin\beta$ and the result now follows easily.

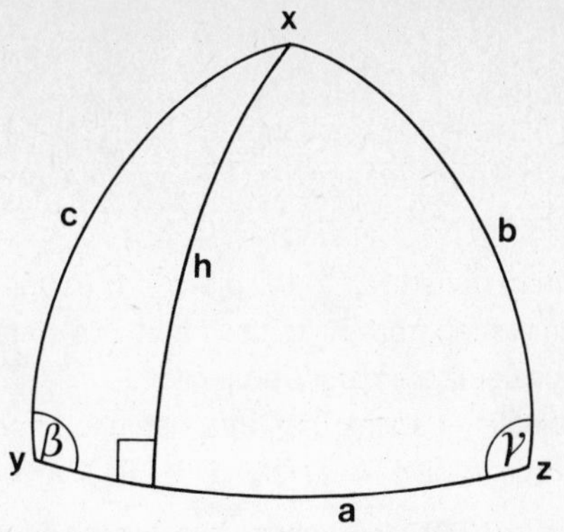


**Note** Euclidean triangles are very close to elliptic triangles with very short sides, so one can recover the usual Euclidean trigonometric formulae by suitable approximations (remember the angles are not small!). For example by putting $\cos \ell = 1 - \ell^2/2$ and $\sin \ell = \ell$ with $\ell = a, b, c$ in the formula ①$_b$ one gets $b^2 = a^2 + c^2 - 2ac\cos\beta$.

**Exercise** Show how the Euclidean cosine rule can be derived from the formulae of hyperbolic trigonometry by a similar approximation.

## Problems

1. Consider a disc $D^2$ made of a material such that the speed of light at a point p is proportional to the Euclidean distance of p from the boundary. Prove that the light rays are the hyperbolic lines of the Poincaré model.

2.* Certain map projections from parts of $S^2$ to the plane are conformal. They can be described as fS where f is a conformal map from $\mathbf{R}^2$ to itself and S is stereographic projection. When $f(z) = -i \log(z)$, show that one gets Mercator's projection.

   A theorem on map projections states that, in a neighbourhood of a point, any two map projections differ, up to first order, by a linear map: $\mathbf{R}^2 \to \mathbf{R}^2$. Prove this.

3.* Show that the co-ordinates in $\mathbf{P}^2(\mathbf{R})$ can be chosen so that the equation of a non-singular conic C is $xz = y^2$. Prove that any elment of PGL(3,**R**) that maps this conic to itself has the form

$$T = \begin{bmatrix} a^2 & 2ab & b^2 \\ ac & ad+bc & bd \\ c^2 & 2cd & d^2 \end{bmatrix}$$

   The map $\xi + i\eta \to [\xi^2+\eta^2:\xi:1]$ defines a homeomorphism $h: H_1 \to H_2$ where $H_1$ is the upper half plane and $H_2$ is the interior of C. If $f \in PSL(2,\mathbf{R})$ is $z \to (az+b)/(cz+d)$, show that $h^{-1}Th = f$. This gives an explicit isomorphism between PSL(2,**R**) and $G_C$. Use this to prove that $G_C$ is the group of all direct isometries of the Beltrami model.

4. The circle of radius r and centre a in a metric space X is the set $\{x \mid d(x,a) = r\}$. If the absolute conic C is a unit circle in $\mathbf{R}^2$, what is the shape of a circle under the Beltrami metric? [First, consider the case where a is the centre of C.] Show that a circle in the upper half plane model is an Euclidean circle but with a different centre.

5. Given $p \in \ell$, let pq be perpendicular to $\ell$. Consider the hyperbolic circle whose centre is at q and touches $\ell$ at p. As q tends to infinity along the perpendicular one gets a limiting curve called a **horocycle** (in Euclidean geometry the horocycle equals $\ell$ but this is not the case in hyperbolic geometry). If the absolute conic is a circle C in $\mathbf{R}^2$ take $\ell$ to be a diameter of C and p to be the centre of C and describe the corresponding horocycle. In the upper half plane model, describe the horocycle when $p = i$ and $\ell$ is the line $x = 0$.

6. If $p \in \ell$, define $d(p,\ell)$ to be d(p,q) where pq is perpendicular to $\ell$. A **hypercycle** is the locus of points equidistant from a line, that is $\{x|d(x,\ell) \text{ is constant}\}$. Find the equations of some hypercycles for the Beltrami and upper half plane models.

7.* Let V be a finite dimensional vector space over **R** and $\varphi$ be a symmetric, non-singular, bilinear form on V. Let $O(\varphi)$ be the subgroup of GL(V) that preserves $\varphi$, show that $\det T = \pm 1$ if $T \in O(\varphi)$. (We have previously considered the Euclidean case, when the signature of $\varphi$ is $\pm$ dim V.) Prove that $O(\varphi)$ is compact if and only if $\varphi$ is definite (the Euclidean case). When $\text{Sign}(\varphi) \neq \pm \dim V$, $v \in V$ is **isotropic** if $\varphi(v,v) = 0$, describe the subset of all the isotropic vectors. $O(\varphi)$ acts on the projective space P(V), how many orbits does it have? If $\text{Sign}(\varphi) = \dim V - 2$, show that $O(\varphi)$ has four components and that if V is two dimensional each component is homeomorphic to **R**. When $\dim V = 4$, $O(\varphi)$ is called the Lorentz group and the set of isotropic vectors form the **light cone** (c.f. relativity).

8.* Take dim V = 2 and use the notation of the previous question. Find a basis such that $\varphi(x,y) = x_1y_2 + x_2y_1$. The isotropic vectors form two lines $\ell_1, \ell_2$; prove that m, n are orthogonal lines with respect to $\varphi$ if and only if the cross-ratio of the four lines is −1.

Prove that $O(\varphi)$ acts on the set of half lines through the origin and that it has four orbits on each of which the action is simply transitive. Show that the matrix $\begin{bmatrix}\lambda & 0\\ 0 & \lambda^{-1}\end{bmatrix}$ that sends $m^1$ to $n^1$ satisfies $(\ell_1\ell_2,m^1n^1) = \lambda^{-2}$. Hence one can define angles by the formula angle $(m^1,n^1) = \frac{1}{2}\log(\ell_1\ell_2,m^1n^1)$. Continue with this approach to obtain a model for the hyperbolic plane.

9.* U(p,q) is the set of non-singular transformations $T: \mathbf{C}^{p+q} \to \mathbf{C}^{p+q}$ that preserve the form

$$z_1\bar{z}_1 + \ldots + z_p\bar{z}_p - z_{p+1}\bar{z}_{p+1} - \ldots - z_{p+q}\bar{z}_{p+q}.$$

Show that U(p,q) is a group; for which p,q is it compact?

Describe the matrices in U(1,1). If T is the matrix $\begin{bmatrix}1 & i\\ i & 1\end{bmatrix}$, show that $T^{-1}AT \in U(1,1)$ if $A \in SL(2,\mathbf{R})$. What is the exact relationship between U(1,1) and $SL(2,\mathbf{R})$?

10. If p, q, r, a, b are points in $P^1(F)$, show that

$$(pq,ab)(qr,ab) = (pr,ab).$$

Let H be the inside of the unit circle C in $\mathbf{R}^2$. If $p \neq q$ are two points in H, let a, b the points where the line pq meets C, define

$$2d(p,q) = |\log(pq,ab)|.$$

If pqr is a triangle, right angled at q, verify that $d(p,r) \geqslant d(p,q)$. Deduce that d is a metric on H. Verify that d is invariant under the group $G_C$.

If H is the unit disc in the plane $x_1 = 1$ in $\mathbf{R}^3$, let $\pi$ denote the projection of H from O onto the sheet $x_1 \geqslant 1$ of the hyperboloid $x_1^2 - x_2^2 - x_3^2 = 1$. Verify that $\pi$ is an isometry.

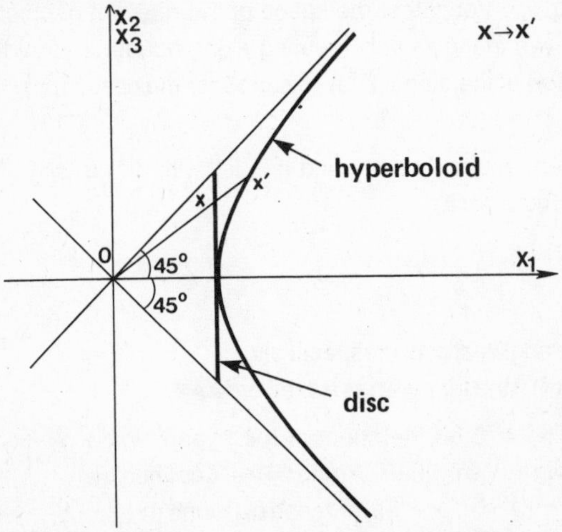


11. Draw pictures of surfaces in $\mathbf{R}^3$ that have negative curvature somewhere. Can they have negative curvature everywhere?

Consider the tractrix

$$x = \operatorname{sech} u, \qquad z = u - \tanh u, \qquad y = 0$$

The tractoid is obtained by revolving the tractrix about the z-axis. What is its curvature? Draw pictures of geodesics on the tractoid.

12. Let xyz be a hyperbolic triangle, right angled at y. It has five quantities associated with it, namely, $\alpha$, $\gamma$, a, b and c. Show that given any two of these quantities, the other three are determined. (There are six essentially different cases to consider.)

13. A hyperbolic triangle has six quantities associated with it, namely, $\alpha$, $\beta$, $\gamma$, a, b, and c. The triangle 'can be solved' from given data if all six quantities can be calculated from the given data. Show that one can solve the triangle if one is given $\alpha$, $\beta$, $\gamma$ as data. From which other triads of the six quantities can the triangle be solved?

14. A right angled pentagon (all angles are $\pi/2$) has edges of length a, b, c, d and e, labelled cyclically. Prove that

$$\sinh a \sinh b = \cosh d.$$

Show that a right angled hexagon can be divided into two right angled pentagons. Deduce a law of sinhs for the hexagon.

15. Consider the points (1,0,0), (cosh u, sinh u, 0) and (cosh v, sinh v$\cos\alpha$, sinh v $\sin\alpha$) on the hyperboloid $x_1^2 - x_2^2 - x_3^2 = 1$ for suitable u, v to prove that there is a hyperbolic triangle whose angles are any given positive numbers $\alpha$, $\beta$, $\gamma$ such that $\alpha + \beta + \gamma < \pi$.

16. An ideal triangle is one such that at least one of its vertices is on the ideal conic C.
    i) If $x \in C$ and $y, z \in H$, find a relation between a, $\beta$ and $\gamma$. What is the area of this triangle?
    ii) If $x, y \in C$ and $z \in H$, can one say anything about $\gamma$?
    iii) If $x, y, z \in C$ and $x^1, y^1, z^1 \in C$, prove that there is an isometry in $G_C$ that sends one triangle to the other.

17. If p, q, r are positive integers, prove that there is a triangle in one of the three plane geometries with angles $\pi/p$, $\pi/q$, $\pi/r$. Verify that the space of the relevant geometry can be covered by such triangles, any two triangles only meeting along edges or at vertices. Such a covering is called a **tesselation** of the plane. Draw pictures for the cases (p,q,r) = (2,3,5), (2,3,6) and (2,3,7).

18. In the upper half plane model, if p = i and $q = ie^d$ with $d > 0$, show that d(p,q) = d. If pqr is a triangle, find the locus of r if
    i) $\alpha = \pi/2$
    ii) $\beta = \pi/2$
    iii)* $\gamma = \pi/2$

    where $\alpha$, $\beta$, $\gamma$ are the angles at p, q, r respectively.

    What is the locus of r if the triangle has a fixed area A?

19. Consider the curve $x_2 = 0$ on the sheet of the hyperboloid $x_1^2 - x_2^2 - x_3^2 = 1$ given by $x_1 \geq 1$, this is a line $\ell$ in H. Verify that the map $f: \mathbf{R} \to \ell$ defined by

$$f(u) = (\cosh u, 0, \sinh u)$$

is an isometry. Deduce that every hyperbolic line is isometric with **R**.

# Further Reading

The best introduction, particularly for the group theoretic approach to Euclidean Geometry, is H. Weyl, 'Symmetry', (Princeton University Press). It is an easy book to read and it deals with symmetry in many contexts.

A classic book containing a wealth of material at a reasonably low level is D. Hilbert and S. Cohn-Vossen, 'Geometry and the Imagination', (Chelsea Publishing Company). It is very readable and its style is discursive.

A text book that covers similar ground to these notes but without assuming as much from other parts of mathematics is H.S.M. Coxeter, 'Introduction to Geometry', (J. Wiley and Sons, Inc.).

A much more comprehensive account written from a similar viewpoint to that taken here is M. Berger, 'Géométrie' (5 vols.), (Cedic/Fernand Nathan).

For an account of the axiomatic approach to geometry at about the level of these notes the reader should read M.J. Greenberg, 'Euclidean and Non-Euclidean Geometries', (W.H. Freeman and Co.).

Other books that contain much relevant material are

L. Fejes Toth, 'Regular Figures', (Pergamon Press).

N.H. Kuiper, 'Linear Algebra and Geometry', (North Holland Publishing Co,).

I.R. Porteous, 'Topological Geometry', (van Nostrand and Cambridge University Press).

H. Eves, 'A Survey of Geometry' (2 vols.), (Allyn and Bacon, Inc.).

Most books on the history of mathematics contain many geometrical topics. A good example is H. Eves, 'An Introduction to the History of Mathematics', (Holt, Reinhart and Winston). A great deal of mathematical insight can be gained by learning about the history of the subject.

There are many other excellent books on geometry. The above are a sample that the reader may wish to look at first.

# List of Symbols

| | |
|---|---|
| $Bij(S)$ | the group of bijections of the set S. |
| $S_n$ | $Bij(S)$ where $S = \{1,2,...,n\}$, the symmetric group. |
| $F^n$ | the vector space of all n-tuples of elements of the field F. |
| $\mathbf{R}^n$ | n-dimensional Euclidean space. |
| $M(n,F)$ | the ring of all $n \times n$ matrices over F. |
| $GL(n,F)$ | the group of all invertible $n \times n$ matrices over F. |
| $SL(n,F)$ | the group of $n \times n$ matrices over F whose determinant is 1. |
| $x.y$ | an inner product on $\mathbf{R}^n$ or $\mathbf{C}^n$. |
| $\Vert\ \Vert$ | the corresponding norm on $\mathbf{R}^n$ or $\mathbf{C}^n$. |
| $d$ | a metric. |
| $O(n)$ | the group of all orthogonal $n \times n$ matrices. |
| $SO(n)$ | the group of orthogonal $n \times n$ matrices with determinant 1. |
| $Aff(X)$ | the affine span of the set X. |
| $T_+(n)$ | the group of upper triangular real $n \times n$ matrices with positive diagonal entries. |
| $S^{n-1}$ | the space of vectors of norm 1 in $\mathbf{R}^n$ (the unit sphere). |
| $R_H$ | reflection in the hyperplane H. |
| $\mathbf{I}(\mathbf{R}^n)$ | the group of all isometries of $\mathbf{R}^n$. |
| $T_a$ | translation by the vector a. |
| $R(a,\alpha)$ | rotation through angle $\alpha$ about the point a. |
| $G(\ell,a)$ | glide by the vector a along the line $\ell$. |
| $I_a$ | inversion in the point a. |
| $M$ | Möbius band. |
| $S(X)$ | the symmetry group of the set X. |
| $D_n$ | the dihedral group with 2n elements. |
| $S_d(X)$ | the rotation group of the set X. |
| $A_n$ | the alternating group on n letters. |
| sign | the sign of a permutation. |
| $J$ | central inversion. |
| $T$ | tetrahedron. |
| $C$ | cube. |
| $D$ | dodecahedron. |
| $\tau$ | the golden ratio. |
| $Stab(x)$ | the stabilizer of x (under a group action). |
| $Orb(x)$ | the orbit of x (under a group action). |
| $L$ | lattice in $\mathbf{R}^n$. |
| $D^n$ | the space of vectors of norm $\leqslant 1$ in $\mathbf{R}^n$, the unit disc. |
| $\mathbf{H}$ | the space of quaternions. |
| $i,j,k$ | unit quaternions. |
| $U(n)$ | the unitary group. |
| $SU(n)$ | the special unitary group. |
| $H_a$ | half turn about the point a. |
| $Aff(\mathbf{R}^n,\mathbf{R}^n)$ | the set of all affine maps $f: \mathbf{R}^n \to \mathbf{R}^n$. |

| | |
|---|---|
| $P^n$ | **real** projective n-dimensional space. |
| **P**(V) | the projective space associated to the vector space V. |
| $\mathbf{P}^n(F)$ | $P(F^{n+1})$, so $P^n = \mathbf{P}^n(\mathbf{R}^{n+1}) = \mathbf{P}^n(\mathbf{R})$. |
| [x:y:z] | homogeneous co-ordinates on $P^2$. |
| V* | Hom(V,F), the space of linear functionals on V. |
| PGL(n,F) | $GL(n,F)/\{\lambda I\}$, (the subgroup $\{\lambda I\}$ is the centre of GL(n,F)). |
| (p,q;r,s) = (pq,ı | the cross-ratio of the points p, q, r, s $\epsilon P^1(F)$, it equals $(p-r)(q-s)/(p-s)(q-r)$. |
| PSL(2,**R**) | the image of SL(2,**R**) $\rightarrow$ GL(2,**R**) $\rightarrow$ PGL(2,**R**). |
| $G_C$ | the group of the Beltrami model whose absolute conic is C. |
| H | the hyperbolic plane. |
| $a^\perp$ | the polar line of the point a. |
| $\ell^\perp$ | the pole of the line $\ell$. |
| O(3′) | the subgroup of GL(3,**R**) that preserves the form $x.y = x_1y_1 - x_2y_2 - x_3y_3$. |

# Index